THE DOMINATION
OF NATURE

THE
DOMINATION
OF
NATURE

William Leiss

George Braziller • New York

Copyright © 1972, by William Leiss
Published simultaneously in Canada by Doubleday Canada, Limited

For information address the publisher:
George Braziller, Inc.
One Park Avenue
New York, N.Y. 10016

Standard Book Number: 0–8076–0646–4
Library of Congress Catalog Card Number: 75–188358

First printing
Printed in the United States of America

For Angela Davis

PREFACE

During the last few years concern with ecology and environmental problems has mounted. Scientists, social thinkers,
public officials, and private groups of citizens have all been
aroused by the possibility of grave crises resulting from the
failure to understand the destructive impact of industrial
society and advanced technologies on the delicate balance of
organic life in the global ecosystem and its various subsys-
tems. The activities which have been initiated range from
international conferences of scientific experts organized by
UNESCO to small community projects on litter, disposable
containers, and polluted streams. Neither the magnitude of
the problem itself nor the nature of the appropriate remedies
is as yet fully apparent. Scientists have recognized that we
understand far too little about the physical processes governing ecological mechanisms and the cycles of self-renewal
through which the balance of organic life is maintained.
Others have called attention to the fact that existing social
institutions do not provide a means for enforcing practical
solutions either in a domestic context (for example, the prac-

tices of industrial corporations) or internationally, as in the case of the pollution of the oceans.

The confrontation of these problems must take place in many ways—community action, scientific research, national and international politics, and critical social thought. This book attempts to make a contribution to the last mentioned of these means. Recent studies, some of them written by scientists long engaged in research in environmental biology and related fields, have stressed the importance of relating certain deeply rooted attitudes toward nature to the new concern with ecological problems. In the present study I have tried to trace the origins, development, and social consequences of an idea whose imprint is found everywhere in modern thought: the idea of the domination of nature. Although its impact has been profound and enduring, this notion has never been studied systematically and critically. I hope that this analysis will provide a better foundation for evaluating the relationship between attitudes toward nature and the ecological crisis. I have not, however, attempted to delineate the precise scope of that relationship, nor have I sought to draw any practical recommendations on the basis of the theoretical investigation presented here.

This book deals with a network of historical tendencies at a high level of generality and abstraction; much further work would be required before its conclusions could be applied to the concrete problems in our contemporary situation. Moreover, I believe that such an effort would be premature. The ecology movement is still in its infancy and as yet is only exploring very tentatively the possible bases for a far-ranging reorientation of attitudes and behavior that would affect the majority of the population. But the emergence of this movement is one of the most hopeful signs of the present.

There are certain inherent advantages and limitations in any purely theoretical study, and the advantages can best be appreciated when the limitations are clearly understood. By isolating aspects of the complex practical activity of everyday life, we may be able, by theoretical reflection, to bare its hidden structural patterns; at the same time, the necessarily abstract character of such work cannot do justice to the rich diversity and sensuous modes of this activity. At best what results can only be an approximation of the social reality: the dynamic of social change always eludes the categories of analysis to some degree and indeed often quickly renders them obsolete. Within these limitations, however, theoretical reflection has an important part to play. By casting the features of these structural patterns into sharp relief, it calls attention to those contradictory elements in human behavior that prevent men and women from achieving their common hopes.

All activity directed toward progressive social change, practical as well as theoretical, takes place in the context of the intractable separation of theory and practice—that is, in the gulf which separates our actual situation from the real promises of human harmony and happiness. No simple formulas, no dogmas derived from our inherited systems of thought, will enable us to bridge that gulf in the near future. Theory and practice occupy different positions along the spectrum of social change and must still grope blindly toward each other. But never before has there been such urgency in this quest. The state of social justice and peace no longer appears as something which men might choose in the ripeness of their wisdom, but as a condition that they must impose upon themselves for fear of destroying all that has been won in human civilization at the cost of so much suffering.

This study of the idea of human domination over nature attempts to explain why the so-called "conquest of nature" can never fulfill the expectations that are conventionally associated with it—generally speaking, the satisfaction of material wants and the establishment of social tranquillity on a universal scale. Thus, paradoxically, such a study appears at first glance to widen the existing gap between theory and practice, since it seeks to demonstrate that certain desired objectives cannot be attained through the kind of activity that is commonly considered to be appropriate for those purposes. But if I have succeeded in arguing the falseness of this conception and in trying to formulate a different outlook on the idea of mastery over nature, the actual result may be just the opposite, for a reorientation of our perspectives may reveal that these objectives are already within our grasp, on the condition that we radically alter the ways in which we pursue them.

Work on this subject has claimed a large portion of my time over a period of four years in as many different countries. Some of the results were first presented under the same title as a doctoral dissertation in philosophy at the University of California, San Diego, in 1969. Parts of this thesis were subsequently published in revised form as a series of articles, together with later writings on related themes, in *The Philosophical Forum, International Social Science Journal, Telos,* and *Canadian Public Administration.* The editors of these journals have permitted me to revise certain sections of the articles for use in this book. Throughout the process of frequent revision my primary objective has been to achieve clarity on a complicated topic without avoiding the difficulties inherent in the subject matter, and during this time my approach has changed more than once, sometimes drastically. I remember the suggestions offered by one of my teachers at the outset, advice which I then stubbornly refused to accept: at the end, after many detours and false starts, I found myself in agreement with him. I hope that he can forgive the ingratitude.

The generosity of many friends in offering encouragement and support could never adequately be acknowledged. In a very real sense this book is as much theirs as mine. I have incurred obligations to them far out of proportion to my ability to repay; no mere listing of names here could reveal how significant that camaraderie in our joint intellectual and political struggles at various places has been for me. The later stages of work on this book were completed much sooner than expected, largely because of the remarkable working conditions established by my colleagues in the Division of Social Sciences at the University of Saskatchewan, Regina Campus, and especially by the members of my department, who generously assumed extra burdens so that I could devote more time to research and writing. The Humanities and Social Sciences Division of The Canada Council awarded me research grants in 1969 and 1971, and during each year of my appointment the Principal's Research Fund at the University of Saskatchewan provided small grants to assist my efforts. I am very grateful for this needed financial support.

All the obligations to my sources of which I am aware are indicated in the footnotes; I apologize to those from whom I have borrowed without realizing it. For the sake of the narrative, references to other works in the text and footnotes are given in abbreviated form. Full bibliographical information for them will be found in the List of Works Cited. Since I have relied so heavily on the contributions of others, I cannot subscribe to the usual disclaimer and pretend that I am alone responsible for whatever faults there are in these chapters: those upon whom I have depended must share part of the blame.

When I last spoke with Angela Davis we were both still graduate students at the University of California, San Diego. At that time she was in the midst of her battle with the regents of the university over her appointment at UCLA, which had added fresh responsibilities to an already overcrowded schedule of teaching, political work in the black community, and the preparation of her doctoral dissertation. Now she is engaged in a far more momentous struggle. To all of these challenges she has responded with a measure of courage, intelligence, and passionate commitment that is a great source of inspiration to others. This book is dedicated to her.

W. L.

Regina, Saskatchewan
September 1971

CONTENTS

PART
ONE

In Pursuit of an Idea:
Historical Perspectives

1

THE CUNNING OF UNREASON

> *The movements of the stars have become clearer; but to the mass of the people the movements of their masters are still incalculable.*
>
> BRECHT, *The Life of Galileo*

1. Recurring Mythologies

In Greek mythology the character of Daedalus combines bold ingenuity in craftsmanship with a restless, amoral disposition. Banned from Athens for the murder of his nephew, whose talents had excited his jealousy, he fled to Crete, where he delighted the royal court with his animated dolls. Having incurred King Minos's displeasure there, he was imprisoned in the Labyrinth, but he was soon free again and with his son Icarus escaped from Crete by fashioning wings of wax and feathers. The reckless Icarus was drowned, but Daedalus continued to mock his adversaries, producing an array of clever devices to celebrate the powers of his boundless, undirected creativity.

In the seventeenth century Francis Bacon turned to the familiar themes of ancient mythology in order to find a medium for his new philosophy, surmising that the cloak of antiquity might render his innovating ideas more acceptable to his contemporaries. In *The Wisdom of the Ancients* (1609), he interpreted the story of Daedalus as a lesson concerning the nature of the mechanical arts, and he emphatically stressed the point that these arts "have an ambiguous or double use, and serve as well to produce as to prevent mischief and destruction; so that their virtue almost destroys or unwinds itself." The tale indeed shows that human society is indebted to mechanical skills for the increase of material provisions and the adornments of culture; but on the other hand "we plainly see how far the business of exquisite poisons, guns, engines of war, and such kind of destructive inventions, exceeds the cruelty and barbarity of the Minotaur itself."

The fable of Icarus failed to impress Bacon: Daedalus's admonition to his son to fly neither too high nor too low seemed but a vulgar reminder to steer a middle course between extremes. But the fable of the Sphinx he found "truly elegant" and instructive, an allegorical representation of science itself. The enigmas the Sphinx proposed resembled the baffling complexities of nature which so far had refused to yield her secrets and her treasures for the improvement of human life. The Sphinx cast her riddles not in the form of idle games, but rather in the context of a life-and-death struggle which impelled men to action. The story of the Sphinx revealed for Bacon the truth that the essence of science is practice. Those who failed of the trial were destroyed, but the successful Oedipus won a kingdom: "All the riddles of Sphinx, therefore, have two conditions annexed, viz: dilaceration to those who do not solve them, and empire to those that do. . . ." [1]

Three centuries later two of Bacon's fellow countrymen took up the mythological themes once again. In 1923, J. B. S. Haldane, biochemist, geneticist, and (with J. D. Bernal and Joseph Needham) member of a remarkable trio of unorthodox British scientists, read to the Heretics' Society at Cambridge University an essay entitled *Daedalus, or Science and the Future*. The shock of the First World War had been felt not only in terms of material devastation but equally in every facet of European consciousness, and one aspect of this reaction was the initiation of a debate on the role of scientific and technological progress in society. Haldane's essay was a contribution to this debate, and his viewpoint still echoes in our own time.

The progress of science, he argued, served as a painful goad which through its social applications forced a reluctant humanity to realize its potential for justice and the satisfaction of needs, both domestically and internationally. He maintained that "the tendency of applied science is to magnify injustices until they become too intolerable to be borne"; in other words, industrial and international injustice ultimately become self-destructive and thus necessitate a more rational and just ordering of social relations. The events of 1914–17 and thereafter constituted a "reductio ad absurdum of war," and it was primarily the achievements of scientists which were responsible for this, since they "enlarged man's power over nature until he was forced by the inexorable logic of facts to form the nucleus of an international government." [2] Haldane permitted himself a few pages of fantasy: he imagined a future society likely to emerge from the tendencies of his own day, a society in which human reproduction takes place by means of artificial conception using a select stock of superior genetic types, and in which all the oceans have turned purple as a result of an alga (deliberately cultivated by artificial fertilizers) that had increased enormously the supply of fish available for consumption.

Almost immediately, Bertrand Russell replied to Haldane with a polemic called *Icarus, or the Future of Science*. The tone of Haldane's pamphlet had not been "optimistic" in the usual sense, but rather had embodied a stoical conviction that the impetus of scientific advance necessarily causes men to improve their social arrangements. Russell, however, countered with a blast of unbridled pessimism. At the outset he announced he was of the opinion that "whether, in the end, science will prove to have been a blessing or a curse to mankind, is to my mind, still a doubtful question." Haldane had in fact been faithful to Bacon's interpretation of the Daedalus myth, for he had stressed the unpleasant or negative effects of scientific and technological progress. But Russell, wishing to broaden the limits of the debate, utilized the story of Icarus far differently. The errant flyer represented not just a particular misjudgment but the fate of an entire civilization which had dedicated itself to the pursuit of scientific progress. The situation in Russell's view was rather simple: "Science has increased man's control over nature, and might therefore be supposed likely to increase his happiness and well-being. This would be the case if men were rational, but in fact they are bundles of passions and instincts." [3]

Russell agreed with Haldane that one of the principal social effects of sustained scientific innovation had been to increase the size of organizational units, particularly economic systems, in accordance with the rules of efficient production. The terminus of this "inexorable logic" would be the political unification of the world. His demurrer concerned the nature of the process by which world-government most likely would be achieved, and to Haldane's innocent fantasy of a biologist he opposed the sober prognosis of a *Realpolitiker:*

> Before very long the technical conditions will exist for organizing the whole world as one producing and consuming unit. If, when that time comes, two rival groups contend for mastery, the victor may be able to introduce that single world-wide organization that is needed to prevent the mutual extermination of civilized nations. The world which would result would be, at first, very different from the dreams of either liberals or socialists; but it might grow less different with the lapse of time. There would be at first economic and political tyranny of the victors, a dread of renewed upheavals,

and therefore a drastic suppression of liberty. But if the first
half-dozen revolts were successfully repressed, the vanquished
would give up hope, and accept the subordinate place assigned
to them by the victors in the great world-trust. . . . Given a
stable world-organization, economic and political, even if, at
first, it rested upon nothing but armed force, the evils which
now threaten civilization would gradually diminish, and a
more thorough democracy than that which now exists might
become possible. I believe that, owing to men's folly, a world-
government will only be established by force, and will therefore be at first cruel and despotic. But I believe that it is
necessary for the preservation of a scientific civilization, and
that, if once realized, it will gradually give rise to the other
conditions of a tolerable existence.[4]

This drastic remedy was recommended by the fact that science
had increased the power of rulers and the ability of men "to
indulge their collective passions" to the point at which the
destruction of civilization itself was a likely possibility. Russell
conjectured that the United States might become powerful
enough to impose its hegemony on the rest of the world and
initiate the gradual evolution toward a tolerable world-government. But, appalled by the logic of his own argument, he
yielded to total despair and concluded with the remark that
"perhaps, in view of the sterility of the Roman Empire, the
collapse of our civilization would in the end be preferable to
this alternative."

In these essays the garb of myth has worn thin: the allusions to the figures of Daedalus and Icarus provide no imaginative "distance," for the same pressing reality which constitutes our everyday experience also appears here. It is a
decisive measure of what has happened during the last two
hundred years that we can no longer think about the future
without estimating in what respect the conditions of human life
will be fundamentally transformed by the achievements of
science and technology. Braced by optimism or pessimism,

anticipating utopia or its opposite, we are compelled to accept the fact that the state of our scientific and technological capability will exert a determining influence on the quality of whatever future is in store for us.

It was not always so. For two thousand years after the appearance of Plato's *Republic* men's expectations of a better order, at least insofar as their earthly ambitions were concerned, were based upon the possibility of radically altering social relationships within the limited framework of a pre-industrial, agriculturally based economy. Whether a "decent" society, characterized by peace, harmony, and the satisfaction of essential human needs for all individuals, was a real possibility under such conditions is of course a debatable question.[5] With the advent of the industrial revolutions there arose the promise of a far more luxurious estate, and gradually the conviction spread that the prospects of maximum leisure and enjoyment were dependent upon sustained scientific and technological progress. To be sure, the older vision never entirely disappeared in modern times: the tradition of utopian speculation down to the present embraces both the paradise of limited consumption, largely restricted to an agricultural economy, and the paradise of expanding needs and satisfactions, tied to an industrial system.[6]

The secular versions of the utopian dream have maintained that the natural environment of the earth contains adequate resources for human happiness and the satisfaction of needs. Human misery arises primarily out of a failure to order social relations justly; given a harmonious society, the arts of men can easily compensate for the material deficiencies inherent in the spontaneous providence of nature. Yet there is a qualitative difference between the preindustrial and the industrial ideals with respect to the degree of human "control" over nature which is considered necessary to insure happiness. In the latter case the possibilities of universal human freedom are explicitly linked with the success of an industrialized productive apparatus based on a technology which has mastered the technique of converting the potentialities of nature to human use in a systematic way. In fact, George Kateb claims that "technology still is that which gives credibility to utopian speculation, that which alone makes interesting and relevant the utopian hope in the twentieth century." [7] The exploitation of the powers of nature, upon which all human art (however "primitive" it may be) depends in some measure, has appeared increasingly important in the social visions of the modern world.

But, in an ironic reversal, the great instruments upon which that exploitation depends—namely, science and technology—have also been classed among the obstacles barring the advent of a new society. To the list of utopia's traditional adversaries, such as war and injustice, have been added the negative features of human relationships arising out of the organization of advanced technologies. Kateb adds: "And it is nothing but the development of technology and the natural sciences that is responsible for the crystallization [of modern antiutopianism] that has taken place." [8] The popular "dystopian" novels of the twentieth century have consistently emphasized these undesirable prospects during the entire period when the most dramatic technological innovations emerged—although one must admit that whatever fears they aroused do not seem to have affected the pace of technological application. Recently much attention has been directed to a new danger posed by these same developments, namely, the threat of ecological disasters. At a 1968 UNESCO conference, a gathering of two hundred scientists concluded that the impact of modern technologies on the natural environment, "if allowed to continue, may produce an extremely critical situation that could seriously harm the present and future welfare of mankind, and become irreversible unless appropriate actions be taken in due time." [9]

The exact nature of this threat to the ecological structure of biological life is as yet unclear. One must also concede that the nightmares of the counterutopian novelists are not necessarily the premonitions of an inevitable future. But the record of the past offers us no justification for underestimating the seriousness of the present situation. Like the sorcerer's apprentice, confident of our mastery over nature we have unleashed incredibly powerful forces and have been caught in the ensuing maelstrom.

The Haldane–Russell exchange is by no means the most extreme formulation of the problem that has vexed social thought and action throughout the twentieth century: How are we to understand and control the social impact of modern science and technology? Philosophy, literature, sociology, history, science fiction, and other intellectual domains have all endeavored to expose the principles governing the hidden dynamic of scientific progress and social development, while under the banners of socialism, technocracy, futurology, and ecology, groups have sought a common ground upon which an institutional fabric capable of containing this dynamic might be erected. Accompanied by an exponential rate of growth in science and technology which has consistently rendered prediction unreliable, these theoretical and practical efforts have barely managed to keep abreast of each new stage of the problem.

In the following pages we will be concerned primarily with the theoretical aspects of the question posed above, that is, with the analysis of some of the ways in which men have represented to themselves the relationship between the accomplishments and the dreams of their sciences on the one hand, and their expectations of social improvement on the other. At times it might seem as if added confusion arises out of any attempt to discuss a contemporary issue in terms of its long historical preparation, especially where the history of ideas is at stake. There is a grain of truth in this impression: concepts both clarify and conceal the nature of the phenomena which they are supposed to represent. For example, the idea of the "natural rights of man" announced the coming of a new political order and simultaneously helped to mask the reality of an economic system characterized by bitter exploitation and class conflict. Similarly, as Russell argues in his essay, men regard science as increasing their control over nature and consequently their happiness and well-being, thereby blinding themselves to the fact that by vastly magnifying their ability to indulge their collective passions they threaten the destruction of civilization. Thus concepts such as the "natural rights of man" and "control over nature" both clarify certain generalized objectives and also inhibit the awareness of fundamental contradictions which thwart the realization of those selfsame objectives.

The problem of understanding and rationally directing the social impact of modern science and technology has so far resisted the analytical power of the received social theories. Bacon could not have known how prescient was his portrayal of science as the Sphinx—allowing some liberty for our own transformation of his intentions, of course. Having developed the material means for the satisfaction of needs so desperately sought in earlier ages, we now find that desires, which are said to be insatiable, can be manipulated to the point where the very concept of human needs is called into question. Even the most industrially advanced nations no longer exercise independent discretion as far as their socioeconomic development is concerned, since an international competition in economic and military activities determines the tempo of technological change. In one of his last essays, Theodor Adorno referred to the "inextricable fatality" which seems to characterize social change at present and which resists precise definition, appearing as an internal concentration of the various dimensions of existence that disguises its true nature under a pervasive "technological veil." [10] This is just the conclusion that emerges from Jacques Ellul's *The Technological Society,* which, for all its faults, illustrates in great detail the tendency of a uniform mode of thinking—valuing above all else the formulation of efficacious techniques for accomplishing whatever tasks happen to be posed—to penetrate all areas of social life in recent times.

Yet we must immediately enter a caveat: as amply demonstrated in the writings of Russell, Ellul, and many others, this subject matter encourages the unfortunate propensity of writers to set forth conclusions at once vague and dramatic. No one remains entirely immune to this practice, despite the most rigorous self-inspection, especially in topics like the present one, where extravagances encountered daily subtly accustom the mind to reasoning loosely. A partial cathartic is offered here in the form of an attempt to analyze carefully the historical, philosophical, and social significance of a crucial conception in the intellectual biography of the modern West: the idea of the mastery of nature.

2. Mastery of Nature—and Man

Found in the most diverse sources from the Renaissance to the present, featured everywhere in recent literature on utopia and on the social consequences of technological progress, the idea of the mastery of nature—a phrase that is used interchangeably with "domination of nature," "control of nature," and "conquest of nature"—presents extreme difficulties for theoretical examination. Often the employment of these phrases is so loose and metaphorical as to render them vacuous; on the other hand, the frequency of their occurrence and the seriousness of the contexts in which they appear hardly permit us to dismiss them as purely literary devices. Moreover, in their lengthy historical career they have acquired many nuances which are usually disregarded, particularly by contemporary authors, because the meaning of these phrases seems so unequivocal at first glance. The resulting confusion is sometimes attributed merely to the use of such expressions, and in accordance with the tactics recommended by the devotees of linguistic analysis in philosophy it is assumed that a vigorous embargo would resolve the matter.

Actually, as we shall see, the terminological vagaries are to some extent visible indicators of hidden contradictions in the social reality. In their dual role of clarifying and concealing this reality, they point to certain connections among diverse historical tendencies and simultaneously obscure or distort other relationships; only a patient restoration of the blurred sections can reveal both the entire mosaic and the actual relations among its component parts. One can understand Aldous Huxley's impatience when he wrote: "It is absurd to attempt—to use that dreadful old-fashioned phrase—to conquer nature." [11] But however dreadful, this expression has represented for many writers a useful way of describing an important modern social phenomenon; and however absurd, this attempt has had unforeseen consequences of profound magnitude. In the opinion of many commentators some of the most paradoxical features of modern society are intimately connected with the notion of "conquering" nature. Some examples will show more precisely the kinds of issues involved.

The domination of nature is regarded as an important part of the modern utopian outlook. The biologist René Dubos remarks: "What is really peculiar to the modern world is the belief that scientific knowledge can be used at will by man to master and exploit nature for his own ends." He adds further that "the direction of scientific effort during the past three centuries, and therefore the whole trend of modern life, has been markedly conditioned by an attitude fostered by the creators of utopias. They fostered the view that nature must be studied not so much to be understood as to be mastered and exploited by man." [12] Paul B. Sears, a botanist and ecologist, sees much the same kind of development: "From the time of Bacon or, to be quite fair, that of Aristotle, scientists have written of the possibilities of a more perfect human society. Of late there has been an increasing emphasis upon the 'conquest' or 'control' of nature as a means to that end." [13] And in a book that describes one of the most famous modern utopias, B. F. Skinner's *Walden Two,* the following expressions are found: "the conquest of nature," "triumph over nature," "scientific conquest of the world," "the urge to control the forces of nature." [14]

Many additional examples of a similar type could be cited. The point is relatively simple: the conquest of nature is accomplished through the agency of modern science as a vital element in the quest for utopia. Yet the images associated with the idea of the conquest of nature have given rise to the conviction that, in the pursuit of this objective, certain countertendencies operate which distort or destroy the character of the utopian dream. This conviction is not to be found solely in the writings of the so-called "romantic" critics of technology, such as Huxley; on the contrary, it is widely shared among writers of different philosophical outlooks and professional specialties.

Some time ago a conservative political theorist, Yves Simon, wrote: ". . . control over natural phenomena gives birth to a craving for the arbitrary manipulation of men; . . . A new lust for domination over men, shaped after the pattern of domination over nature, had developed in technique-minded men." [15] More recently, Robert Boguslaw, who devoted a book to arguing that computer technicians represent the authentic utopian planners of our day, concluded his work as follows:

> Our own utopian renaissance receives its very impetus from a desire to extend the mastery of man over nature. Its greatest vigor stems from a dissatisfaction with the limitations of man's existing control over his physical environment. Its greatest threat consists precisely in its potential as a means for extending the control of man over man.[16]

The puzzling affinity of these two trends has also been noted in a study which attempts to trace the historical development of modern utopian thought from its beginnings in the sixteenth century. The author notes: "Both the utopists and the scientist Newton lived in the formative period of a concept of progress based on the conquest of nature, that is, science. But somehow this concept has led to the conquest of man, too, in the utopian societies of Orwell and Huxley." [17] Finally, in a speech delivered to the UNESCO scientific congress mentioned earlier, the Director for Science and Technology of the United Nations Department for Economic and Social Affairs voiced similar sentiments in the context of a growing concern among scientists about the matter of worldwide environmental destruction:

> In recent centuries, however, the world has been increasingly dominated by a dualistic world-view in which the distinction between man and his environment has been particularly stressed. This view accepts as a virtual axiom that man's foremost task consists in the progressive establishment of complete mastery over all of non-human nature. But, in recent times, man has tended to become so dominant on earth that he is now approaching a position where he constitutes one of the principal aspects of his own environment and in which environmental mastery would require the subjugation even of human nature by man. [18]

These are representative samples from an extensive literature. In them, as in the larger body of material from which they are drawn, there is widespread agreement on the following points: (1) the effort to master and control nature has an essential connection with the modern utopian vision; (2) the mastery of nature is achieved by means of scientific and technological progress; (3) the attempt to master external nature has a close and perhaps inextricable relationship with the evolution of new means for exercising domination over men—or, alternatively, human activity becomes so much a part of the natural environment that mastery of nature and mastery of man are only two aspects of the same process. There is a virtual unanimity on the first two points which is lacking with respect to the last, but this seems to result more from the subjective inclinations of individual authors than from a convincing demonstration of a particular viewpoint.

In the context of modern utopian thought, "mastery of nature" is a shorthand expression for the guarantee of an adequate material provision for human wants. The burgeoning requirements of individuals apparently can only be met by a highly developed productive apparatus which continually locates new resources in nature and transforms them into desired commodities. The more sanguine authorities are convinced that a level of abundance sufficiently high to satiate the whims of the most demanding populace (presumably on a worldwide scale) is a genuine practical possibility based on present forecasts of technological capabilities. If the material desires of men and women are expected to expand infinitely, then mastery of nature in this sense means an ongoing search for adequate sources of satisfaction, and under certain conditions every level of attainment might meet an escalating discontent arising out of appetites accustomed to regular stimulation.

The second point—the idea that the conquest of nature is achieved through the accomplishments of the modern natural sciences and their concomitant technology—is agreed upon almost universally. Very few writers pause to ask whether this judgment implies that the domination of nature is the explicit guiding purpose of the sciences or a more or less accidental by-product of their discoveries. Of course this is not at all surprising, since the link between mastery of nature as the outcome of scientific progress and as increasing satisfaction of human wants seems to be self-evident: the research laboratories and the mass-production assembly lines are often two departments of the same enterprise and demonstrate this link as an everyday occurrence. Normally both aspects are amalgamated in the conventional studies on this topic, such as R. J. Forbes's *The Conquest of Nature: Technology and its Consequences*. But these two frames of reference for the mastery of nature may well be incongruous, and it is still necessary to determine the precise sense in which science and technology constitute a "conquest" of nature.

On the third point there is little agreement. Although the fond hopes of earlier epochs have not been extinguished, unease about the future now demands recognition; the transition "from utopia to dystopia" is one of the major literary preoccupations of our time,[19] for the attempted conquest of nature almost inevitably seems to result in frightful new means for the exercise of domination in human affairs. The same scientific and technological order which promises to liberate mankind from its universal enemies (hunger, disease, and exhausting labor) also enables ruling elites to increase their ability to control individual behavior. In the imaginations of dystopian novelists such as Huxley, Orwell, and Zamiatin, the dangers of the familiar forms of despotism pale beside the prospect of things to come: the subjects of earlier tyrannies recognized their slavery in the overt controls which restricted their physical movements and in the terror which the minions of authority inspired in them, whereas the citizens of the future, manipulated at the very sources of their being, will love their servitude and call it freedom.

Perhaps no one really believes that this is likely to occur. But even among the ordinary ranks of social analysts, who are normally not given to fictional excesses, some will concede that the novelists have drawn attention to a genuine dilemma in contemporary society. Like Boguslaw, for example, who was quoted on this point earlier, they are content to acknowledge it only in passing, although on the basis of their own arguments it would appear to deserve more direct attention. One must ask: *why* is there apparently a connection between the conquest of nature and the "conquest of man"? Is it inevitable that the scientific and technological instruments utilized in the domination of nature should produce a qualitative transformation in the mechanisms of social despotism?

The attempt to clarify these issues must proceed in the face of the most diverse attitudes toward the social processes that accompany the attempted human control of nature. The range of attitudes is bounded at each extreme by increasingly irreconcilable conceptions of the idea of mastery over nature. The following passage illustrates the view of the matter which simply ignores all disturbing elements:

> Man's relationship with the renewing elements in his natural environment is at an important stage in history. It appears that a total control of nature is possible in a not very distant future. Many ecologists deny this, claiming that nature is too complex to be reflected in the simulation of any computer technology. Such assertions indicate that we have not yet managed to describe nature completely. Until the subject has been fully described, it is not likely that the controller can freely manipulate it with a superiority to natural processes. But the means of acquiring that complete description are already well developed, as are the economic and social conditions that make a greater control of nature necessary.[20]

Here the human control of nature is represented as a purely technical problem, since the author has abstracted from the element of *interaction* between man and nature: "nature" is a fixed object, a sphere of pure externality, a stage-set for the display of human activity. Only the barest hint of another interconnecting dynamic is contained in the allusion to social factors, and the author makes no further mention of it. At the other extreme, a psychologist who regards "man's growing control over nature" as the most important revolution of our time suggests—apparently in all seriousness—that the silence of extraterrestrial space may imply that other species became extinct "by using their control of nature to destroy their life." [21]

More to the point is the judgment of René Dubos, a scientist who has devoted many years to the study of human health and development in relation to environmental biology. Dubos believes that the rising level of wasteful consumption in the industrialized nations is inherently self-destructive; in his opinion, a better knowledge of ecology will reveal "that using the power produced by scientific technology for mastering nature" as we do at present "is not sufficient to sustain civilization, not even industrial civilization." [22] The crucial question for him concerns the *quality* of the civilization that we wish to have. And part of that problem consists in determining more clearly what is meant by the domination of nature.

3. Reason and Unreason

The juxtaposition of quotations in the preceding pages was not intended merely to serve as a demonstration that "mastery of nature" and its substitutes are favored phrases among contemporary writers, nor was the objective simply to document the resulting confusions. These phrases have been used in serious attempts to grasp the essential interrelationships among real historical phenomena, among which the following are the most significant: the growing scientific understanding of the "laws of nature"; the continued success in turning scientific discovery into technological innovation at an ever more rapid pace; the ability, won in the industrial revolution, to apply technological innovation to the production of goods on a mass basis; and the hope that all of these tendencies would greatly reduce or even eliminate the familiar sources of human misery and social disorder. The surprising frequency of these usages indicates that they have so far served as valuable guides for interpreting this ensemble of events. However, the persistent failure to appreciate not only what is illuminated by these concepts, but what has been so well concealed, makes necessary a thorough study and critique of them.

For the most part, neither the positive nor the negative features of the historical situation which emerged from the converging developments listed above were the result of explicit designs. Most men labor in the service of immediate interests and problems, theoretical or practical, and their own interpretations of the significance of their activity are often spurned by later generations, which invariably regard themselves as more self-critical. Moreover, enlightened practices often appear on the historical stage encased in a barbarous husk; the most unlikely individuals and groups, who are fated to be the unwitting vessels of progress in civilization, sacrifice their earthly existence to the cause of a hidden rationality which offers them no compensation. Hegel proposed a famous notion, the "cunning of reason," in order to call attention to this bizarre aspect of human history, wherein reason "sets the passions to work for itself, while that which develops its existence through such impulsion pays the penalty, and suffers loss." [23]

Mastery of nature at first glance appears to be an excellent illustration of Hegel's idea. The scientific, technological, social, and industrial changes which taken together comprise what is usually meant by the conquest of nature are the outcome of innumerable individual and group motivations and impulses that remained largely unrecognized by the participants. Psychological, religious, and philosophical elements coexist in this network with economic and political factors, and the union has never been entirely harmonious; at times it was the sinister requirements of warfare, for example, which most effectively repressed the discordances and supplied the necessary stimulus for important breakthroughs. Yet from this welter of conflicting aims and interests emerged a qualitative transformation in the human ability to exploit the forces of nature.

There is, however, another side to the matter that merits

more attention than it has received thus far: with due apologies to Hegel we may refer to it as "the cunning of unreason." This expression may serve as a guide to the unifying theme in the succeeding pages, the general perspective of which can be briefly described as follows. Early in the seventeenth century there arose a belief in the possibility of formulating a radically new method to guide scientific investigation. The chief propagandists for this idea, Bacon and Descartes, who were regarded by their contemporaries as the twin prophets of a new age for mankind, argued that by accepting this novel method men would achieve "mastery over nature." This formula encompassed two distinct thoughts: (1) the new method would permit an explanation of natural phenomena far superior to what obtained in their day with respect to such criteria as generality, consistency, and conceptual rigor; (2) the fruits of the method also would consist in *social* benefits—notably an increased supply of goods and a general liberation of the intellect from superstition and irrationality—that would enable men to control their desires and to pursue their mutual concerns more justly and humanely.

Thus the rationality of the new scientific methodology was considered to be an independent force capable of leavening the social milieu. Throughout modern times the hope has persisted that this rationality would take hold in ever wider domains and thereby improve the condition of social relationships. Its enduring message is nicely epitomized in a recent statement: "The hope is that scientific knowledge, as the mode of adaptation and of human control over nature, may also be a major instrument of human self-control, by means of rational intelligence." [24] This is an accurate representation of a conviction, alternately waxing and waning during the last three hundred years, that has remained a principal article of faith in modern Western civilization. That hope has been consistently frustrated, but the reasons why this is so have never been explained adequately. Social development continues to defy all attempts at rational control and is governed instead by the puppetry of a hidden dynamic—the cunning of unreason, whose most fateful manifestation is the process whereby the rationalism of modern science and technology becomes caught in the web of irrational social contradictions. We must attempt to see what part the concept of mastery over nature has played in that process.

Marcuse has said that "man's struggle with Nature is increasingly a struggle with his society." [25] Perhaps a better formulation would be that the social struggles among men gradually absorb the struggle of men with nature, or, in other words, that the instruments through which men transform the resources of nature into means for the satisfaction of desires are regarded more and more as a crucial object of *political* conflict, both domestically and internationally. For example, a concern with ecology necessarily becomes part of a *social* movement because the problem of reversing the present self-destructive treatment of the environment cannot be separated from that of challenging the authoritarian decision-making powers vested in corporate and governmental institutions. Similarly, in international affairs the life-and-death competition between capitalism and socialism sets the frenzied pace of scientific and technological innovation and thus subjects "mastery over nature" to the rule of an uncontrolled dynamic which contains disastrous implications.

This integration of the relationship between man and nature into the sphere of social conflict necessitates a fresh approach to the interpretation of the mastery of nature. The focus of the analysis must be directed at the questions that have never been adequately treated: What is the basis for the presumed link between the "control" of nature as the outcome of scientific knowledge and the "self-control" of human behavior? Does the predominant conception of mastery over nature provide an adequate framework for the interpretation of the historical trends to which it calls attention? And if not, is there a better conception?

A consideration of these questions will demonstrate the correctness of Lukács's remark that nature is a "social category." [26] Mastery of nature is not simply the objective of scientific research, but rather a profoundly influential image of certain expectations and meanings that successive generations have attached to their scientific enterprise; and it is only through an examination of this complex of meanings that we can find the key to the social significance of that enterprise. To date there has been very little critical analysis of this subject. (A significant exception is Marcuse's distinction between two kinds of mastery over nature, "repressive" and "liberating." [27] This is an important step, but he has not yet developed the distinction adequately.) In the following chapters I have tried to clarify the historical origins of the idea of mastery over nature and to answer the questions posed above.

Were it possible to penetrate the multiform guises of the problem and to distill a single statement of it from the vast literature in which mastery of nature figures so prominently, we might phrase it as follows: the cunning of unreason is revealed in the persistent illusion that *the undertaking known as "the mastery of nature" is itself mastered.*

2

MYTHICAL, RELIGIOUS, AND PHILOSOPHICAL ROOTS

Let us establish a chaste and lawful marriage between Mind and Nature, with the divine mercy as bridewoman. And let us pray God, the Father of men and nature as well as of lights and consolations, by Whose power and will these things are done, that from that marriage may issue, not monsters of the imagination, but a race of heroes to subdue and extinguish such monsters, that is to say, wholesome and useful inventions to war against our human necessities and, so far as may be, to bring relief therefrom.

BACON, *The Refutation of Philosophies*

1. The Demonic Smith

Several versions of the following story are found in Zulu legends. In accordance with the prevailing customs a woman brought her newborn baby to the two-headed talking birds who bestowed names on all children. This child was hideously deformed and the birds, discerning the presence of terrible evil in it, announced in alarm that the child would have to be destroyed at once. The mother rebelled against the judgment and escaped into the jungle, where she remained hiding in a

25

deep cave, raising her child in the bowels of the earth. One day the mother noticed that under the young boy's fixed gaze iron ore began to melt. Out of the ore he fashioned a flying metal robot with powerful teeth and jaws, and when his mother begged him to destroy his creation he instead ordered the robot to kill her. Then he created an army of similar robots and, emerging from the cave, led his forces against the two-headed birds, utterly destroying them and conquering the world. He established himself as supreme ruler, and all the world's people lived in perfect comfort, since the innumerable mechanical creatures acted as their slaves and performed every task. Gradually men and women lost their ability to do anything at all, even reproduce their kind; so the robots created semihuman zombies out of animal flesh which were capable of reproduction.

The Tree of Life and the earth goddess were appalled at this interference with their sacred functions, however, and resolved to destroy the offenders. They sent a huge flood, but the ruler ordered the robots to build gigantic rafts upon which the many survivors could live permanently, while he and his court reveled in even greater luxury than before. Since the sun was continually hidden by storm clouds, the ruler directed the construction of an artificial one, which in the people's opinion shone more brilliantly than had the real sun. But finally the goddess and the Tree of Life marshaled overwhelming force and obliterated all traces of this evil civilization.

Like the myth of Daedalus and many related legends in different cultures, this story preserves the ambivalent emotions of fear and desire aroused by the instruments (especially those formed of metals) which men have devised to improve their earthly lot. Examples of this mythological tradition are compared and analyzed in a book written by the well-known ethnologist Mircea Eliade, *The Forge and the Crucible*. Eliade shows that in widely scattered ancient cultures there is a uni-

form feature, namely, "the sacredness of metal and consequently the ambivalent, eccentric and mysterious character of
all mining and metallurgical operations." [1] In some cultures
certain aspects of metallurgical activity, for example the
smelting operation, required human sacrifices. Everywhere
rites of purification, especially involving sexual taboos, were
associated with mining because the inner earth was considered
sacred to various spirits or deities; religious ceremonies accompanied the opening of new mines in Europe until the end
of the Middle Ages.

The invention and fabrication of tools out of various metals
always had magical or divine overtones, and the smith was
often regarded as a magician because he supposedly possessed
the secret formulas that enabled him to make things appear
which were not "natural." His creations had a sacred character which persisted also in the work performed by means of
them, for "in the symbols and rites accompanying metallurgical operations there comes into being the idea of an
active collaboration of man and nature, perhaps even the
belief that man, by his own work, is capable of superseding
the processes of nature." [2] The feeling of power and independence in human activity and the attitude of superiority
to nature that is engendered by the use of tools is paralleled
by recurring fears that these instruments are "demonic"; thus
ceremonial rituals are required in order to placate the spirits
and to "humanize" the instruments through which men have
altered the natural order.

The surviving remnants of these once-powerful beliefs are
not only to be found among the so-called "primitive" tribes
which still inhabit certain areas of the globe. Both the researches of psychoanalysis and the experience of fascism have
demonstrated how archaic impulses coexist and interact with
the structures of modern rationalism. Any perusal of the
enormously popular science fiction literature will quickly
reveal that the impact of advanced technology among the

"civilized" populations in the industrialized countries releases deeply hidden fears—normally repressed in public life—concerning man's ability to control the techniques and forces placed at his disposal by science. In fact, there are in this literature innumerable contemporary analogues of the Zulu legend related above, together with countless variations on the theme of the enslavement of the human race by its erstwhile mechanical servants.

Of course this does not mean that archaic mental patterns have remained perfectly intact from the time of their origins. The point is that, despite the undoubtedly superior rationality of modern techniques, *the instruments and procedures of science and technology have by no means lost their magical aura* in either preindustrial or industrialized societies. The average citizen understands very little of their logical foundations or working principles and plays a purely passive role (as consumer) as each attained stage is transcended by a newer one at an accelerating rate. The transistor radio, for example, has been accepted by many peoples whose autochthonous technology is enveloped in ancient rhythms and primitive implements; but although the same radio clashes less with their immediate surroundings, the ordinary inhabitants of the industrialized regions have exercised just as little discretion concerning its "appearance" as did their less opulent neighbors.

Thus one significant cause of the enduring archaic component in the reaction to modern technology is that a magical quality still clings to it. The suppressed terrors which it evokes arise in part out of a dim awareness on the part of many individuals that apparently no one really "controls" the increasingly sophisticated instruments devised for the control of nature; as a result these instruments seem to possess an independent dynamism which continually threatens to overawe the human agencies surrounding them. At the same time, the obvious material benefits obtained by these means and the bewildering complexity of the whole institutional apparatus of modern society paralyze the will to call up into consciousness these hidden fears, to confront and possibly to surmount them. And so the inner desire and terror that characterize humanity's oldest experiences with technology continue to feed a kind of fatalism whereby people gratefully accept the fruits of human ingenuity while dreading the eruptions of uncontrollable malevolence from its handiwork.

2. *The Lord of the Earth*

In a fascinating lecture delivered at a meeting of the American Association for the Advancement of Science in 1966, the historian Lynn White argued that "the present increasing disruption of the global environment is the product of a dynamic technology and science which . . . cannot be understood historically apart from distinctive attitudes toward nature which are deeply grounded in Christian dogma." [3] He tried to show that an inherited religious outlook had been attached to the growing scientific and technological competence in Western civilization and had largely determined the way in which this competence was applied in the service of human wants. Thus, according to White, the roots of the present ecologic crisis are religious in nature, and a religious transformation, rather than technological remedies, would be required to solve it.

Whether or not we agree with his conclusions, we should recognize that White has identified a critical aspect of the relationship between man and nature, for in addition to the mythical residue discussed above, our religious heritage remains a vital source for interpreting the intellectual mosaic of mastery over nature. As White points out, until the eighteenth century almost every great scientist was also preoccupied with religious problems, and it is reasonable to conclude that their theological concerns influenced their conception of the *sig-*

nificance of scientific progress. Together with their less accomplished contemporaries, they shared a distinctively Christian attitude toward nature which was based on certain familiar teachings about man's role in the creation and which had guided Christianity's triumph over pagan animism.

A common feature of the religions that dominated the ancient world was the belief that all natural objects and places possessed "spirits." These had to be honored in order to insure oneself against harm, and before appropriating natural objects for human use man was required to placate the spirits through gifts and ceremonies. The Judaeo–Christian religion, however, maintained that "spirit" was separate from nature and ruled over it from without; it also taught that to some extent man shared God's transcendence of nature. Only man of all earthly things possessed spirit, and thus he did not have to fear the resistance of an opposing will in nature; the Bible seemed to indicate that the earth was designed to serve man's ends exclusively. White concludes that "by destroying pagan animism, Christianity made it possible to exploit nature in a mood of indifference to the feelings of natural objects." [4]

Obviously, this argument does not attempt to account for the novel philosophical, methodological, and experimental principles which characterize modern science, since the hegemony of Christianity spans a much wider historical period. One rather unorthodox philosopher has indeed claimed that Christianity prepared the "ontological" ground for modern science: it contradicted the Greek view that the earth lacked the perfection of the heavenly bodies, thus permitting the eventual formulation of the idea that matter is everywhere "equal," that is, that the same mathematically expressible laws govern the behavior of all matter in the universe.[5] We are concerned here not with the internal development of science itself, however, but rather with the teachings that define man's relationship to nature. These are of course only one aspect of Christian doctrine, and the significance of any such part in the lives of individuals varies with time. Yet Christianity still retained a powerful hold on the European mind during the formative years of modern technology and science; the received imagery of their faith provided the protagonists of the new science with some ready-made categories through which to interpret their accomplishments. Science conceived as the winning of mastery over nature seemed to be the natural fulfillment of the Biblical promise that man should be lord of the earth.

The creation story in the Book of Genesis announces the sovereignty of God over the universe and the derivative authority of man over the living creatures on the earth. The Biblical account emphasizes the possession of absolute power as the basis of sovereignty, and it is just this element of power which essentially separates man from other created things: "In the idea of a covenant between man and lower creation, man is distinguished not by the possession of spiritual faculties, but by being lord of the universe under God, Who is Supreme Lord." For "man does not rule over the animal kingdom because he is God's image: rather, he is God's image precisely because he rules over the animal kingdom, thus sharing God's universal dominion." [6] The decisive question for Christian theological commentary on this point was how the Fall affected man's dominion on earth. The existence of wild animals was regarded as evidence that there had been a partial loss of authority on account of sin, for it was assumed that in the Garden of Eden all animals obeyed man's bidding. The domestication or destruction of the wild animals would be a sign that the earthly paradise had been restored. The legends recounting the deeds of the early saints who retired into the wilderness all speak of their accomplishments in taming beasts as proof that they were reasserting the rightful human sovereignty enjoyed prior to the Fall.[7]

St. Thomas Aquinas's views on this point probably represent the most carefully considered Christian statement. In the state of innocence man exercised his authority over the animals by "commanding" them, but his mastership over plants and inanimate objects "consisted not in commanding or in changing them, but in making use of them without hindrance." In this interpretation the Fall did not essentially alter man's relationship with the nonanimal world; rather, its major consequence was that these things were no longer accessible "without hindrance" and had to be appropriated by labor. A partial control over the animal kingdom, that is, over various domestic animals, was retained. Aquinas stresses the point that man "excels" all animals not by virtue of power but rather because of his reason and intelligence, and it is in respect to these latter attributes "that man is said to be according to the image of God." [8]

To be sure, Aquinas's fine distinctions were often ignored, and for most Christian thinkers it was sufficient to emphasize the point that man's sovereignty on earth had not been lost (at least not entirely) as a result of sin. Martin Luther's *Commentaries on Genesis,* for example, explicitly concur with this general outlook.[9] The idea that man stands apart from nature and rightfully exercises a kind of authority over the natural world was thus a prominent feature of the doctrine that has dominated the ethical consciousness of Western civilization. There is no more important original source for the idea of mastery over nature; Bacon's formulation of the idea, for example, which more than anything else is responsible for its wide modern currency, is saturated in the Christian tradition. This does not mean that every age has understood the idea in the same way or has been enamored of it to the same degree. Hannah Arendt correctly remarks that in the Old Testament man is given command over all living creatures, not over the earth itself; but it is not strictly true that "the notion of man as lord of the earth is characteristic of the modern age"—that is, *only* of the modern age—and that this notion contradicts "the spirit of the Bible." [10] There is ample evidence to show that in premodern times as well the Biblical passages gave rise to the view of man as lord of the earth. The important point is that in different ages the same conceptual form is filled with a different content. Whatever may have been the understanding of it in earlier epochs, the modern version of this conception is firmly associated with the ongoing successes in science and technology.

The continuity of the Christian tradition thus provided a consistent image of man as lord of the earth based on the Biblical creation story. The most significant aspect of this imagery is the degree to which the religious setting in Genesis has always been interpreted in *political* terms. The God of Genesis is pictured as the absolute ruler of the universe who has delegated subordinate authority to man for the management of affairs on earth. The Hellenistic philosopher Philo Judaeus wrote in *On the Creation:* "So the Creator made man all things, as a sort of driver and pilot, to drive and steer the things on earth, and charged him with the care of animals and plants, like a governor subordinate to the chief and great King." And in his book *The Primitive Origination of Mankind* (1677) the English jurist Matthew Hale declared that "in relation therefore to this inferior World of Brutes and Vegetables, the End of Man's Creation was, that he should be the Vice–Roy of the great God of Heaven and Earth in this inferior World." [11] These are two examples from a vast flood of religiously inspired literature in which the Old Testament myth is invested with political connotations. The Biblical setting of the relationship between man and the rest of creation was made comprehensible to countless generations by means of a political analogy. Therefore it is no accident that the terms which concern us in this book—mastery, domination, and conquest—all spring from the same source.

One might be tempted to dismiss the whole matter as a case of false analogy. On this view "mastery of nature" would be explained simply as an expression which embodies an illicit transference of meaning; in other words, it is a case of terms which when properly used only designate relationships among men being carelessly employed in describing the interaction between man and nature. In many cases such an approach might be correct, but not I think in the present instance. Here we are faced not with a short-lived aberration, but rather with a remarkably consistent pattern of thought that has roots deep in the main cultural traditions of Western society. It is a collective vision, not an individual thinker's idiosyncrasy, and mistaken or not it still constitutes the mold of great hopes. The modern notion of mastery over nature carries the living force of this image, particularly the meaning bestowed on it by Christianity, into the present; to dismiss it as a terminological error is to deprive oneself of a valuable clue to the social consciousness of our time.

The discussion in the following pages attempts to follow this clue in various ways; here only one of its implications will be noted. The Biblical portrayal of man as lord of the earth is part of a larger canvas in which the political analogy is mediated and transcended. Man's will is not the highest principle in heaven or on earth, but instead is checked and limited by ethical norms established independently of it. Similarly, the surrounding world of nature has a purpose entirely apart from its function as the material basis of human activity: it is a divine creation and therefore sacred. Nature has a double aspect. In its immediate presence, as the source of satisfaction for vital human needs, it necessarily arouses utilitarian modes of behavior (which may differ widely in structure and detail); reflectively, however, nature appears as the visible testimony of God's providence and thus must be regarded from the perspective of its value as an aid in understanding the divine scheme. Lynn White's argument must be qualified to this extent, namely, that Christian doctrine sought to restrain man's earthly ambitions by holding him accountable for his conduct to a higher authority.

So long as Christianity remained a vital social force in Western civilization, the notion of man as lord of the earth was interpreted in the context of a wider ethical framework. Religion's declining fortunes, however, led to the gradual secularization of this notion in imperceptible stages, and in contemporary usage it reveals few traces of its Judaeo–Christian background. The identification of mastery over nature with the results of scientific and technological progress, in connection with the cultural antagonism of science and religion in the eighteenth and nineteenth centuries, dissolved the traditional framework. For Francis Bacon there was no apparent contradiction between his religion and his hopes for science—in fact the image of man as the lord of nature clearly helped him to unite the two; but the Baconian synthesis, so characteristic of the seventeenth century, has not endured. The purely secular version of this image retains the various associations derived from the political analogy discussed above while shedding the ethical covering that both sustained and inhibited it. In its latter-day guise, mastery over nature loses the element of tension resulting from the opposing poles of domination and subordination in the religiously based version and adopts a unidimensional character—the extension of human "power" in the world.

3. The Philosophical Monkey

Renaissance illustrations, for example in the works of the English mathematician and mystic Robert Fludd, often represent human art and knowledge by the figure of an ape. The idea of human ingenuity as the "ape of nature," that is, the imitator of nature's operations, is much older, but the Renaissance brought this image to new prominence as part of its innovating intellectual accomplishments.[12] The Renaissance witnessed the first steps in an intense upsurge of interest in the workings of nature, in its "secrets" and its "treasures."

Exuberant speculative flights were transformed into cosmological systems, and at the same time a technological literature based on the craft of skilled artisans began to spread. These vastly dissimilar enterprises possessed a common root, namely, a growing confidence in the human ability to expand the limits of the control and utilization of natural forces. No one doubts any longer that during the Middle Ages there was a steady succession of quite remarkable technological innovations. There is also little question that the rate of development began to accelerate at the end of this period and that, perhaps more importantly, a new consciousness—displaying a greater relative appreciation, even a fascination, for a theory and practice that promised to extend man's powers—took hold in wide areas of Europe.

Quite appropriately, many contemporary scholars have claimed that the Renaissance is the primary modern source for the idea of mastery over nature. Various elements of the idea have been found in the Renaissance theory of natural magic, in its alchemy, cosmology, and astrology, in its vision of the Magus, and even in Machiavelli's precepts on the "mastery of fortune" in political life. To some extent this genealogy is indeed accurate, as the summary to be drawn from these sources will show. But it is essential to emphasize at the outset the fragmentary and diffuse nature of these references. They do not form an integrated pattern, nor can they rival the historical consistency of the Christian contribution discussed earlier. They have largely been forgotten or neglected in recent times. Nonetheless these diverse Renaissance currents, however bizarre some of them seem to us now, can all claim some share both in the formation of this idea and in preparing the intellectual ground for its wide reception.

A profound revaluation of man's status is of course one of the essential features of Renaissance philosophy. The *Platonic Theology* of Marsilio Ficino boldly announced that man "imitates all the works of the divine nature, and perfects, corrects and improves the works of lower nature. Therefore the power of man is almost similar to that of the divine nature, for man acts in this way through himself." [13] For the philosophers the efficacy of man's powers was demonstrated clearly in the practice of magic. This subject had to be handled very delicately, however, because the Church strongly condemned magical exercises; their device consisted in a somewhat tenuous distinction between good and bad, or natural and demonic, magic. Ficino's tortuous reasoning, undertaken in an effort to overcome his Saturnian horoscope with the aid of more favorable stars, established a foundation for this distinction, and from these innocent beginnings there arose an irresistible interest in magic among the learned everywhere in Europe. [14]

Ficino's brand of natural magic was intended solely for use within closed aristocratic circles and it aimed at purely personal effects, namely, the improvement of an individual's fortune through purification of the soul. [15] It relied on the *vis verborum,* the "power of words," according to the theory that by discovering the true names of natural things, such as planetary bodies, we can command the forces possessed by those things to our benefit. Another kind, named "transitive" magic, was practiced as a means of controlling animate beings (both men and animals) by influencing their imaginative faculties. Perhaps no one put more faith in the practical effects of magic than the Italian mystic Tommaso Campanella, who tried to use it to win over the possessors of temporal power, kings and Pope alike, to his scheme for the unification of the world.

Paolo Rossi has shown that the Renaissance writers made magic more acceptable by contending that it never violates the principles of nature; on the contrary, they maintained, magic is the "servant of nature," nature's imitator and assistant, and its attempted operations remain faithful to the natural order. Agrippa von Nettesheim's *Of the Uncertainty and Vanity of the Sciences* (1527) was explicit:

> Therefore natural magic is that which having contemplated
> the virtues of all natural and celestial things and carefully
> studied their order proceeds to make known the hidden and
> secret powers of nature. . . . For this reason magicians are
> like careful explorers of nature only directing what nature
> has formerly prepared. . . . Therefore those who believe the
> operations of magic to be above or against nature are mis-
> taken because they are only derived from nature and in
> harmony with it.[16]

The partisans of magic claimed that the apparent miracles
and tricks accomplished by the adept only seemed to be so
because others did not understand the reasons or causes of
what was produced; actually the magician only "anticipated"
and assisted the outcome of natural operations. Clearly these
principles, although intermixed with much fantastic lore,
helped prepare the intellectual climate in which science later
flourished.

These themes crystallized in the idea of the Magus, the
archetypical magician of whose many incarnations the Renaissance preferred the legendary Hermes Trismegistus.[17]
The Magus typified all that was noble and pure in the realm
of magic; he was represented as a learned and dignified
philosopher who had nothing in common with the ordinary
despised necromancer. The sources of his art were Neopla-
tonic astrology and "practical Cabala," the Hebrew variation
of magic, but it was argued that none of this was at all in-
compatible with the Christian religion. The philosopher Pico
della Mirandola was chiefly responsible for reintroducing the
figure of the Magus in this form. Distinguishing demonic
magic from natural philosophy ("good" magic), Pico argued
that "even as the former makes man the bound slave of
wicked powers, so does the latter make him their ruler and
their lord"; by means of the latter the Magus weds "earth to
heaven, that is, he weds lower things to the endowments and

powers of higher things." [18]

A fascination with numbers was an integral part of Renaissance magic. Numbers were regarded as providing the key to unlock the secrets and infinite powers of nature, and much mathematical genius was expended in the search for the correct formula. So powerful was the attraction of this idea that mathematical mysticism and mathematical science remained associated to some degree until the nineteenth century; the Renaissance stimulus clearly prepared the ground for the possibility of conceiving an exact science of nature in mathematical terms. The Englishman John Dee (1527–1608), an outstanding mathematician who passionately studied magic in a search for the secrets of nature, is one of the best examples of a thinker caught in these overlapping trends. Ernst Cassirer regards Kepler as marking a decisive turning point, for Kepler expressly demarcated for himself the separate pursuits of astronomy and cosmological speculation. [19]

In her book *Giordano Bruno and the Hermetic Tradition,* Frances Yates contends that the Renaissance conception of the Magus was a crucial and necessary step in the formation of modern thought. In the figure of the Magus she detects a "psychological reorientation" of the will toward action, a broad change in attitude which asserted the dignity of man's operational abilities in the mundane world. [20] Toward the end of her book she relates this point to the problem of explaining the rise of a new science in the seventeenth century. This is not a matter concerning the developmental stages of a different methodology, but rather the prior question of locating the sources whence sprang an intense burst of interest in nature's workings and which nourished a determination to understand the operational principles of the world.

This argument runs the risk of infinite regress: what in turn caused such a collective reorientation of the will? There is no profit in pursuing causal chains here, and only a flexible scheme of interacting and intersecting factors—socioeconomic, psychological, cultural, political—is appropriate. Insofar as the intellectual aspects of this network are concerned, there are of course many continuities linking the Middle Ages with the succeeding centuries; but Yates's point (which is stressed also by other authorities, notably Alexandre Koyré) that a new collective attitude toward the relationship of man and nature arose in the Renaissance is certainly correct. The literature of this period reveals a steadily growing fascination with the "secrets" and "operations" of nature and a desire to penetrate them in order to possess power and riches, so much so that already in the seventeenth century some authors decried what had become in their view a fetishistic pursuit.[21] The extravagant dress in which the Renaissance thinkers often clothed their formulations of the anticipated human mastery over nature offends modern sensibilities, and some recent commentators fail to acknowledge these disreputable entries as genuine forebears of the idea. Yet although the genealogical lines are suspect in particular cases, taken together the Renaissance notions constitute a recognizable portion of the conceptual inheritance on which contemporary usage lives.

4. Faustian Synthesis

"The alchemist," wrote Paracelsus, the great practitioner of this trade, "is he who helps to develop to the extreme limits intended by nature that which nature produces for the benefit of mankind." [22] All of the various strands which we have been examining, the archaic, the religious, and the philosophical, were woven together in the heroic figures of the alchemist and his wonder-working literary cousin, Doctor Faustus. In them the new craving to unveil and master the hidden forces of nature displays strongly obsessive characteristics: the pathology of the alchemist has become a rich area for psychological research. The Faust legend, which originated in the late sixteenth century and became extraordinarily popular, obviously struck a powerful sympathetic chord, one which still resounds today. Thomas Mann's novel, like Goethe's play so much more complex than the average presentation of the story, preserves in part not only the structure but even the actual language of the original version.[23] A fundamental ambiguity dominates both the legend and the response to it, for the notion of an irresistible lust to command nature's secret energies arouses the terrifying, complementary fear and guilt represented in the diabolical pact.

The professional alchemist's actual fate was far less dramatic. He had only his soaring visions with which to compensate himself for the unrewarding drudgery of a lifetime spent in the presence of foul-smelling substances. The following passage written by one of their most famous number, Agrippa von Nettesheim, is not untypical:

> But who can give soul to an image, life to stone, metal, wood or wax? And who can make children of Abraham come out of stones? . . . No one has such powers but he who has cohabited with the elements, vanquished nature, mounted higher than the heavens, elevating himself above the angels to the archetype itself, with whom he then becomes co-operator and can do all things.[24]

Their contemporaries probably discounted such pretensions while at the same time awaiting eagerly the outcome of their researches. Elizabeth's ministers hoped to finance a fleet for the Spanish campaign with gold wrought by transmutation in the laboratory, and toward the end of the sixteenth century the Emperor Rudolph II assembled at his court in Prague astrologers and alchemists from every corner in Europe so that they might conduct an organized search for the key to nature's secrets, the philosopher's stone.

Jung attempted to analyze the alchemical texts with reference to the theory of projection. The alchemists were operating with the merest fragments of knowledge about matter and its structural principles, and yet they were convinced that they had only to discover a few operational tricks in order to gain the long-sought prize: complete mastery of nature, that is, the ability to duplicate the work of creation. They were pathetically unaware even of how vast was the darkness in which they groped; the intensity of their desire together with the poverty of their knowledge, Jung claims, resulted in their projecting the contents of the unconscious as the properties of matter. Eliade subsequently modified this theory by relating certain themes in the alchemical writings to universally held archaic beliefs.[25] For example, the notion of the "torture of metals" which occurs frequently in these writings and is the basis of the transmutation experiments, reveals definite affinities to the general pattern of primitive initiation rites, especially the aspects of dismemberment and rebirth in a different form. According to Eliade the alchemists "projected on to matter the initiatory function of suffering."

These archaic survivals blended with another strong impulse which had been a common feature of the intermingling religious and philosophical traditions: the striving for individual perfection. Alchemy deliberately favored an arcane language and an elaborate symbolism which epitomized its inner-directed intentions and its conviction that the way to perfection was revealed only to a select few. For the alchemist the goals of mastering nature and of bringing to perfection his own nature were identical. The vision of interlocking moral and scientific objectives was succinctly stated by Newton, as reported by a friend who heard him praise the alchemists in his old age: "They who search after the Philosopher's Stone by their own rules obliged to a strict and religious life. That study fruitful of experiments."[26]

What makes the figure of the alchemist directly relevant to the present study is the argument, advanced separately and in different contexts by Jung and Eliade, that a thoroughly secular version of the alchemist's dream has become the driving spirit of the modern age. Jung remarks (somewhat over-dramatically) that "alchemy was the dawn of the scientific age, when the daemon of the scientific spirit compelled the forces of nature to serve man to an extent that had never been known before," and he adds that the psychic sources of the will to conquer the world are most apparent in the alchemical literature. Eliade argues that the concepts of material progress and the conquest of nature through science and technology crystallized in the nineteenth century and became the basic social ideology which still persists today. "It is in the specific dogma of the nineteenth century," he writes, "according to which man's true mission is to transform and improve upon Nature and become her master, that we must look for the authentic continuation of the alchemist's dream." Thus in Eliade's view the alchemists "anticipated what is in fact the essence of the ideology of the modern world." [27]

These sweeping judgments must be tempered with caution if the kernel of truth that they possess is to be extracted. The "ideology" of the present is not quite so uniform as it might appear at first glance from the perspective of the hoary legends resurrected by ethnology and depth psychology. The genuine common interest in material progress cannot conceal the substantial actual and potential alternatives which confront us, and to slight them is to make nonsense of much contemporary history. In fact, the more triumphant is the ideology, the more necessary is the task of choosing among alternative modes of realizing the expectations it arouses. For neither the dogma of material progress nor that of mastery over nature really has a determinate, fixed content. And if only to escape the blind compulsiveness of social forces which become ever more threatening, men are required to formulate more rationally the objectives underlying these dogmas. Eliade's contention that "capitalist, liberal and Marxist" ideologies are only variations of the universal dogma of material progress, which is sufficiently confusing in itself, misses the main point: in any variation there will be contradictory features (not the same ones, of course) which must be transcended.

The kernel of truth in the argument is the idea that archaic impulses persist in social behavior and in the unconscious; in other words, in terms of the framework just outlined, these impulses play a part in the interaction of conflicting factors. In his solitary search for the techniques of mastering nature the Renaissance alchemist tapped those ancient resources; traversing the web of associations found in the rich hermetic and magical literature, he wove the mythical, religious, and philosophical strands into an incredible tapestry whose borders overlap here and there with modern science and technology. Our idea of mastery over nature was formed against this background by Bacon and others, and the traces of their struggle to remove its diabolical overtones are still discernible. When the conviction spreads that the heralded conquest of nature is actually being accomplished through science and technology, then the surviving elements of the alchemist's psychological drama are externalized in some measure and are reflected in social attitudes, for example in the kinds of fears mentioned at the beginning of this chapter.

One basic point emerges from an inquiry into the historical roots of the idea of mastery over nature: this idea has long been immersed in the darker side of the human psyche and has retained associations with evil, guilt, and fear even in its recent secularized form. Whatever the reasons for this may be, it is essential at least to recognize the fact and to evaluate its significance, for only then does the possibility arise that the archaic response to technology might be finally transcended. Unfortunately one usually finds this problem treated only in the most superficial discussions of technology and society, in books composed entirely of platitudes and clichés but spiced with allusions to Frankenstein, whereas the conventional references to the domination of nature rarely go beyond a few well-worn Baconian slogans which by now have lost their original vitality. In this we do Bacon—and ourselves—a disservice.

3

FRANCIS BACON

> *[Bacon] was an old man when he was allowed to leave prison and return to his estate. His body was weakened by the efforts he had spent to bring about other people's ruin and by the sufferings other people had inflicted when they ruined him. But no sooner did he reach home than he plunged into the most intensive study of the natural sciences. He had failed in mastering men. Now he dedicated his remaining strength to discovering how best mankind could win mastery over the forces of nature.*
>
> BRECHT, *Tales from the Calendar*

1. Introduction

The vicissitudes of Bacon's reputation as a thinker are in themselves an index to the changing historical circumstances of the period since his death.[1] In the first flush of excitement over the demonstrated successes of scientific experimentation he was universally praised as the herald of a new order and was christened "the secretary of nature." During the last half of the seventeenth century in England no philosopher past or present was regarded as his equal, and his fame soon spread to the Continent. The Enlightenment celebrated him as its chief source of inspiration: d'Alembert extolled his "sublime

45

genius" in the *Preliminary Discourse to the Encyclopedia* and Diderot's prospectus for that work stated that its "principal debt" was to Bacon. Even those who damned him contributed to the recognition of his influence, as in the case of the leading philosopher of the conservative reaction to the French Revolution, Joseph de Maistre, who traced all the evils of the age back to the English Lord Chancellor.

Bacon aimed both at changing the prevailing cultural and philosophical attitudes and equally at effecting drastic institutional reforms. The great idea which possessed him all his life was that of *organized* scientific research; he drafted many different proposals during his years in government service before sketching in his old age the vision which was to cast a spell over all succeeding generations, namely, that of a research establishment for science and technology—called Solomon's House or the "College of the Six Days' Works"—which is described in his *New Atlantis* (1627). Few radical innovators have met with such a prompt response: the times were ripe. In the fifty years following Bacon's death educational reformers influenced by the *New Atlantis* began organizing technical schools to promote instruction in the mechanical arts. The Royal Society (chartered in 1662), whose founders explicitly acknowledged the inspiration of Bacon's lifelong dream, was soon joined by the *Académie des Sciences* in France and by similar bodies formed elsewhere in Europe.

During the nineteenth century, however, despite the balanced and favorable treatment of Bacon's thought that could still be found in such sources as Hegel's and Feuerbach's histories of philosophy, the clamor against him began to grow. This was primarily the result of an increasing interest in the methodology of the sciences and in the philosophical justification of scientific principles, for in respect to these concerns Bacon's epistemology was found wanting. A reaction set in which can be detected in places even today; according to his critics, Bacon's heart was in the right place, but unfortunately his actual prescriptions for a new scientific approach were a hopeless muddle. Of course there is much truth in this view, as any contemporary reader of Bacon will soon discover. Yet it overlooks the fact that the thinkers of the seventeenth and eighteenth centuries had admired him not for his epistemology —which they largely ignored—but for his passionate advocacy of the need for sustained progress in the mechanical arts and the physical sciences.

Beginning with Cassirer's masterly reassessment of modern epistemology since the Renaissance published early in this century, our understanding of Bacon's thought in its historical context has steadily improved. Cassirer showed that Bacon had taken over and strengthened the great Renaissance redirection of interest toward the natural world, the revaluation of philosophical categories and approaches that encouraged the careful observation of empirical physical phenomena. Subsequent interpretations, especially the studies by Paolo Rossi, have deepened this perspective by exploring the detailed connections among Bacon's attitudes toward myth, religion, magic, and philosophy, and by showing how his conceptions of science and progress relate to that larger framework. The subtle character of Bacon's indebtedness to magic and alchemy, as well as the nature of his critique of those sources, have only recently been appreciated, and in large measure this explains why the Baconian conception of mastery over nature has always been treated so superficially.

This has been a strange fate indeed, since no expressions in the history of modern thought have been quoted and praised more frequently than Bacon's pithy formulations of this idea. To be sure, one cannot assume that the domination of nature is an idea which struck Bacon's readers as forcefully in earlier periods as it did later. On the other hand, the concept of mastery over nature has been regarded as an outstanding contribution of Bacon's world-view for a long time. To cite only one instance, here is Ludwig Feuerbach's summary of the Baconian doctrine from his *History of Modern Philosophy from Bacon of Verulam to Benedict Spinoza* (1833): "Natural science has therefore no other goal than to more firmly establish and extend the power and domination of men over nature. But the domination of men over nature rests solely on art and knowledge." [2] Modern scholars readily acknowledge the concept of mastery over nature as a fundamental element in Bacon's philosophy, and contemporary literature is full of quotations and references to it. Unquestionably Bacon's writings are the single most important direct source of present usage. Yet the fact remains that his concept of the mastery of nature, however familiar it might be, has not been analyzed critically. This task thus forms the necessary prelude to any evaluation of its contemporary significance.

The brilliance of Bacon's prose—its limpidity, grace, and muscular imagery—is partly responsible for his readers' failure to examine his statements closely. After three-and-a-half centuries his phrases still retain an aura of excitement. He possessed the ability to encapsulate complex thoughts in brief, well-turned expressions which must be carefully unraveled so that the full range of their associations may become apparent.

2. The Recovery of the Divine Bequest

Bacon's great achievement was to formulate the concept of human mastery over nature much more clearly than had been done previously and to assign it a prominent place among men's concerns. Its dangerous connection with the megalomaniacal delusions of the alchemists was severed; and, still infused with Renaissance energies, it was wedded once again by Bacon to the predominant cultural force of that time, namely, Christianity. The idea was made "respectable." Of course the notion of man's dominion over the earth had always been a part of the Judaeo–Christian heritage, as we have seen;

but in the context of the emerging constellation of historical factors at that time—the economic, social, political, scientific, and technological changes which capitalism fused together into a system of expanding productivity—this notion took on a wholly new significance. The precise way in which Bacon reformulated it was crucial, for Christianity's hold on the European consciousness remained strong even as the traditional social basis of organized religion was being eroded by capitalism. Bacon provided the formula whereby the idea of mastery over nature became widely acceptable, a formula which also was easily secularized as the cultural impact of religion gradually diminished.

In Bacon's view religion and science were engaged in a mutual effort to compensate for the damage incurred as a result of the expulsion from Paradise: "For man by the fall fell at the same time from his state of innocency and from his dominion over creation. Both of these losses however can even in this life be in some part repaired; the former by religion and faith, the latter by arts and sciences." [3] All the subsequent stages of his argument are dependent first upon this original distinction between the two different consequences of sin, loss of moral innocence and loss of dominion, and second upon the claim that two separate agencies (religion and science) are appointed to mitigate the attendant evils. This distinction enabled Bacon to maintain that the attempt to master nature through scientific progress did not violate God's plan; on the contrary, it was precisely by means of these arduous steps that the divine judgment, "In the sweat of thy brow shalt thou eat thy bread," was fulfilled.

Bacon perceived that behind the reluctance of society to encourage scientific innovation was the fear that man might incur God's wrath by interfering with the natural order of things. Thus he took great pains to stress the "innocence" of the scientific endeavor. In a little stage-play he devised as a young man for Elizabeth, the first of his many unsuccessful efforts to elicit royal support for organized research, he suggested that of all the exercises of the mind, "the most innocent and meriting conquest" was "the conquest of the works of nature." [4] Similarly, in one of his earliest important essays he wrote that men's discoveries in the arts and sciences were the outcome of a game which they played with God,

> as if the divine nature enjoyed the kindly innocence of such hide-and-seek, hiding only in order to be found, and with characteristic indulgence desired the human mind to join Him in this sport. And indeed it is this glory of discovery that is the true ornament of mankind. In contrast with civil business it never harmed any man, never burdened a conscience with remorse. Its blessing and reward is without ruin, wrong or wretchedness to any. For light is in itself pure and innocent; it may be wrongly used, but cannot in its nature be defiled. [5]

How easily in this extraordinary passage Bacon glides from one context to another! The blamelessness of human artistry in a religious sense is contrasted with the apparently unavoidable wrongs suffered in social activity, but the two are juxtaposed in such a way that we are encouraged to believe that the arts and sciences *preserve* their innocence while serving as instruments of social progress. The archaic fears about the demonic character of human craftsmanship are put to rest: the means of restoring man's rightful dominion over the earth are beyond good and evil because this dominion itself originates in the general state of premoral innocence which characterized human life before the Fall.

The religious frame of reference provides the guarantee that the proposed qualitative expansion of the arts and sciences will not lead to uncontrollable upheavals: "Only let the human race recover that right over nature which belongs to it by divine bequest, and let power be given it; the exercise thereof will be governed by sound reason and true religion." [6]

Is this only a pious hope? A careless misjudgment on Bacon's part? I do not think so, for it seems to me that his concept of the human mastery over nature is too deeply immersed in a religious context for us to assume that he was only paying lip service to conventional morality. Certainly one cannot overlook his passionate concern with material progress, since he constantly reiterated the point that his reformation in natural philosophy was intended to "increase and multiply the revenues and possessions of man." But throughout his writings from the beginning to the end of his career mastery of nature is always referred to a larger religious scheme.

His favorite image for conveying the sense of restored dominion over nature which would be realized through scientific progress is Adam's naming of the animal species. Here the magical and religious heritages coincide, for as we have seen the protagonists of magic and alchemy believed that power over nature could be won by discovering the "true" names of natural objects, that is, by means of the *vis verborum*. Bacon argued that the objective of human knowledge is "a restitution and reinvesting (in great part) of man to the sovereignty and power (for whensoever he shall be able to call the creatures by their true names he shall again command them) which he had in his first state of creation." [7] Against the alchemists' intellectual arrogance he insisted that one's first duty in any inquiry is to respect what is divine. Having made this much clear, he then cautioned his fellow men to take care

> that in flying from this evil they fall not into the opposite error, which they will surely do if they think that the inquisition of nature is in any part interdicted or forbidden. For it was not that pure and uncorrupted natural knowledge whereby Adam gave names to the creatures according to their propriety, which gave occasion to the fall. It was the ambitious and proud desire of moral knowledge to judge of good and evil, to the end that man may revolt from God and give

laws to himself, which was the form and manner of the temptation.[8]

This clear separation of natural knowledge and moral knowledge gradually became a cardinal principle of modern thought: it echoes in the fashionable contemporary distinction between "facts" and "values," according to which questions of values constitute a unique discourse outside the scope of "scientific" knowledge. Bacon establishes it through his interpretation of the Fall and his related contention that no amount of knowledge concerning nature's operations will improve our understanding of God's plan. Nature resembles the artisan's finished product, which reveals the power and skill of its maker but not his image.[9] By these means Bacon (and others like Galileo and Descartes) sought to allay any fears that the scientific investigation of nature would shake the grounds of faith. But there is a hidden dichotomy in his argument. On the one hand, the legitimacy of mastering nature—and the "moral innocence" of the arts and sciences through which it is accomplished—is derived from a divinely ordained relation between man and nature; in other words, Bacon justifies the human mastery over nature by referring it to the condition of man before the Fall, when man himself existed in a state of moral innocence and in perfect harmony with the whole of creation. On the other hand, the recovery of dominion over the earth through the arts and sciences in no way helps to restore the state of innocence, for that is a separate problem of moral knowledge and faith which is in the domain of religion.

The result of this dichotomy is that Bacon—together with virtually all of his readers—failed to notice the necessity of demonstrating one crucial aspect of his argument, namely, that it is through the progress of the arts and sciences that "mastery of nature" is achieved. Indeed this point seems so self-evident to us that by raising it as a problem one risks being considered slightly odd. But it is only because the full context of Bacon's famous phrases have so rarely been appreciated that it appears strange to request such a demonstration. Why is the recovery of the divine bequest not the result of moral progress rather than scientific progress? This will not seem to be such an empty question if we recall the legend of the early saints in the wilderness mentioned in Chapter Two: it was their exemplary moral life, not their superior scientific knowledge, which was believed to be the basis of their restoration of that dominion over the animals possessed by Adam.

Bacon retained the familiar imagery but altered its content decisively, and his subtle transfusion is one of the outstanding moments in modern intellectual history. By casting his plea for scientific progress in a familiar religious mold he managed to win wide acceptance for a novel conception of mastery over nature, and at the same time he unwittingly charted a course for later generations which led to the gradual secularization of this idea. His contention that science shared with religion the burden of restoring man's lost excellence helped create the climate in which earthly hopes flourished at the expense of heavenly ones. More important even than this, however, was his coupling of innocence and dominion. Bacon claimed to have identified a way back to the latter—through science—which was quite different from the means available for regaining the former (of course in neither case would there be a complete recovery). But was he also hinting that the one might carry the other in its train?

Certainly he never said as much, and most likely he would have regarded such a suggestion as blasphemous. However, I believe that there may be grounds for assuming that in some form this idea was subconsciously present. In a passage quoted earlier we are told that the "blessing and reward" of discoveries in the arts and sciences "is without ruin, wrong or wretchedness to any," and that the "light" which is manifested in these discoveries, that is, their contribution to the amelioration of human existence, "is itself pure and innocent." These selfsame innovations are the promised instruments of mastery over nature, which if systematically developed will yield "all operations and possibilities of operations from immortality (if it were possible) to the meanest mechanical practice." [10] This highly irreligious slip, which is hardly rectified by the parenthetical remark following it, is a clue to the subconscious associations in Bacon's mind. The victory over death would be a sure "sign" (a favorite Baconian expression) of innocence restored *as a result of* a dominion regained through science.

There are few such clues, and the overt evidence of Bacon's writings does not allow us to draw any firm conclusions on this point. It is only in the *New Atlantis* that the concrete ramifications of this problem are apparent, and we shall return to it in the course of examining that work. No matter what importance we assign to this veiled element in Bacon's thought, however, we cannot fail to recognize the fact that throughout the succeeding centuries variations of this theme have frequently recurred. The most common formulation was mentioned at the end of Chapter One: the rationality of science and technology works as an independent force to infuse rationality into the social process as a whole. In other words, mastery of nature through scientific and technological development is conceived as a means of *social* progress, as if the "innocence" of the instrument could be conferred on the larger context in which it functions. But we can no longer believe, as Bacon could, that within this larger context religion would govern the exercise of the "right over nature" which belongs to the human race "by divine bequest," that is, that an extrahuman dimension would preserve the original link uniting dominion and innocence. When the concept of mastery over nature is thoroughly secularized, the ethical limitations implicit in the pact between God and man, whereby the human race was granted a partial dominion over the earth, lose their efficacy. The religious casing in which Bacon had embedded his plea for a fresh conquest of nature failed, but the idea itself emerged intact and in secular dress fired the imagination of later periods.

In Chapter Two we noticed that the theology of Genesis relating to the creation myth has always been interpreted with the aid of political metaphors. Bacon's efforts are no exception to this rule. On the contrary, together with his religious pre-occupations his ambitious labors in politics and law naturally caused him to inject new vigor into these metaphors. Referring to the sciences as a "suit" at law, he proclaimed: ". . . I mean (according to the practice in civil causes) in this great Plea or Suit granted by the divine favour and providence (whereby the human race seeks to recover its right over nature), to examine nature herself and the arts upon interrogatories." [11] The coercive overtones of the word "interrogation" are rein-forced in the delightful Baconian phrase, the "inquisition of nature," which appears in a passage quoted earlier. Changing his analogies somewhat, Bacon elsewhere urged government support for scientific research by reminding his audience that "as secretaries and emissaries of princes are allowed to bring in bills of expenses for their diligence in exploring and un-ravelling plots and civil secrets, so the searchers and spies of nature must have their expenses paid. . . ." [12]

The kingdom of nature is like any other realm, subject to conquest by those who command the requisite forces. Bacon decries the unwillingness of men to study diligently the mun-dane phenomena of nature, for "most certain it is that he who will not attend to things like these, as being too paltry and minute, can neither win the kingdom of nature nor govern it." [13] And what is the form of government appropriate to this political metaphor? The image is one of despotism, although we cannot tell whether or not it is the benevolent type: "I am come in very truth leading to you Nature with all her children to bind her to your service and make her your slave." [14] How much emphasis should be placed on his choice of imagery is perhaps a matter of taste. Nevertheless this sentence offers some insight into the meaning of mastery over nature for Bacon, since it is at least possible to conceive of nature as man's helpmate rather than as his slave. The difference lies not in the degree to which human desire succeeds in finding gratification, but in the condition of desire itself. The appropriate corollary to the image of the helpmate is a sense of inner responsibility in the expression of desire, that is, the recognition of the necessity for subjecting human wants to rational control, whereas Bacon's image suggests that nature is the object of a will which is itself enslaved by uninhibited and uncontrollable desire.

The conquest of nature promises liberation from the "inconveniences of man's estate," in other words, relief from the adverse conditions of existence which arise out of the prevailing state of the relations between man and nature. Bacon's chief concern was to convince his audience that the growth of knowledge would alter that relationship so that steady material progress could be expected. Another dimension of his thought that is not often noted, however, in his contention that "learning doth make the minds of men gentle, generous, maniable, and pliant to government; whereas ignorance makes them churlish, thwart, and mutinous." Thus there are two interrelated advantages brought by the advancement of learning, namely, relief from the inconveniences caused by the relation of man and nature together with those resulting from the relation of man to man. Bacon says:

> Neither is certainly that other merit of learning, in repressing the inconveniencies which grow from man to man, much inferior to the former, of relieving the necessities which arise from nature. . . . [For men] are full of savage and unreclaimed desires, of profit, or lust, of revenge, which as long as they give ear to precepts, to laws, to religion, sweetly touched with eloquence and persuasion of books, of sermons, of harangues, so long is society and peace maintained; but

if these instruments be silent, or that sedition and tumult make them not audible, all things dissolve into anarchy and confusion.[15]

Here, in a proto-Hobbesian perspective, mastery of external nature is related to mastery of internal nature (human nature). The growth of knowledge will not only enlarge the bounds that determine the degree of satisfaction of material wants, but will also serve as an instrument for repressing the permanent instinctual threat to social peace. Burgeoning desire will be gratified via the achievements of science and technology, and its destructive potentialities will be checked by the devices of culture.

3. The Inquisition of Nature

Bacon's vision of the social context most advantageous for the conquest of nature is set forth only in his *New Atlantis*. Before turning to that essay we must examine some of the language employed in his description of the process by which nature is to be mastered. In order to avoid a direct confrontation with the received philosophical truths of his day, which would have augmented (he thought) the vain and fruitless debates he sought to circumvent, Bacon claimed that he was proposing objectives for knowledge radically different from those prevailing in the schools. He maintained that what distinguished the opposing view from his own was not its falsehood but simply the fact that it was guided by different goals:

For the end which this science of mine proposes is the invention not of arguments but of arts; not of things in accordance with principles, but of principles themselves; not of probable reasons, but of designations and directions for works. And as the intention is different, so accordingly is the effect; the effect of the one being to overcome an opponent in argument, of the other to command nature in action.[16]

His maneuver is quite apparent, as he fully intended it to be.

In his opinion the natural philosophy of his time was worthless because barren of operational results. Its sophistical dialectic was well suited to "civil business" but not nearly "subtle" enough for the investigation of nature. Clearly the refined and delicate logic of the universities was too fragile to meet the Baconian demand for a natural philosophy that would encompass two related operations, "the one searching into the bowels of nature, the other shaping nature as on an anvil." [17] Bacon thought that the chief obstacle to a reformation of knowledge was a deeply rooted psychological condition. The persistent failure of the accepted natural philosophy to improve man's ability to utilize the powers of nature bred an attitude of despair, and this attitude in turn inhibited the exploration of new approaches. This despair had to be routed if there were to be any possibility of making nature "serve the business and conveniences of man" to a far greater extent. To this end Bacon proposed a new outlook toward the relationship between nature and human art. According to the traditional philosophical doctrines, art merely supplements nature, finishing or perhaps slightly correcting what nature has to offer; but we err gravely, says Bacon, in thinking that art has "no power to make radical changes, and shake her [nature] in the foundations; an opinion which has brought a great deal of despair into human concerns." His language is stunningly explicit on this point: the mechanical inventions of recent years do not "merely exert a gentle guidance over nature's course; they have the power to conquer and subdue her, to shake her to her foundations." [18]

He proposed a new attitude, one based on confidence and hope rather than on despair, one which stakes everything on the "victory of art over nature" in the contest of the two. A fateful new stage in the struggle between man and nature is to be initiated. He insists again and again that great rewards will follow if we carefully observe a simple principle: we command nature by obeying her. This famous formula is often interpreted in such a way as to soften the tone of the language employed by Bacon elsewhere, for example in the passages cited just above. Some commentators have claimed that it sounds a note of humility in man's attitude toward nature. But this interpretation is somewhat forced and invents inconsistencies which do not really exist in Bacon's work. A simpler explanation is available. According to this principle human art takes on the connotation of artfulness or cleverness, that is, it adapts itself to the necessities inherent in things. By following this formula we learn to set traps for nature in the form of experiments, carefully observing her workings and deducing the particular steps involved; then, having discovered the course leading to the end result, we are able to duplicate the process at will. Bacon summarized the outlines of this method in memorable terms: "For you have but to follow and as it were hound nature in her wanderings, and you will be able, when you like, to lead and drive her afterwards to the same place again." [19]

We discover how nature performs her operations by observing her under three conditions. First, there is nature "free and at large," which is the state wherein nature runs unhindered in her course, as in the regular movements of the heavenly bodies and the reproduction of animal and vegetable life on earth. Second, there are the "errors" of nature that result in deformed creatures. Third, there is nature in "bonds," transformed by human art; and "seeing that the nature of things betrays itself more readily under the vexations of art than in its natural freedom," this last condition is the most auspicious of all for the increase of scientific knowledge. The experiments in the mechanical arts deal with nature "under constraint and vexed; that is to say, when by art and the hand of man she is forced out of her natural state, and squeezed and moulded." [20] Bacon's use of political metaphor in his effort to dramatize the idea of mastery over nature even pervades this description of scientific procedure. Human art and knowledge are the weapons with which men compel nature to do their bidding.

How much Bacon's striking manner of conceptualizing mastery of nature influenced his readers is impossible to estimate. Certainly contemporary usage, even when it pays homage to Bacon's inspiration, seems rather listless by comparison. The psychological dynamic of mastery over nature is still discernible to some extent in Bacon's language, offering some clues to the network of psychic associations in which the desire for "power" over nature was enmeshed. The vital legacy of magic and alchemy is revealed in his terminology, which displays strong overtones of aggression (including the sexual aggression connected with the feminine gender of the noun and the use of "her" as the pronoun): "hounding," "vexing," and "subduing" nature.[21] This legacy and the psychological complex which inspired it is no longer visibly present in the references to mastery of nature found in modern social theory, although traces of it still appear in the less respectable science-fiction genre.

Freud interpreted this dynamic in terms of the opposition of Eros and Thanatos: "The instinct of destruction, when tempered and harnessed (as it were, inhibited in its aim) and directed toward objects, is compelled to provide the ego with satisfaction of its needs and with domination over nature."[22] He never explicity discussed any further the concept of mastery of nature, however, and the general index to his collected works does not list a single additional use of this phrase. But Norman O. Brown emphasized it in his interpretation of Freud's work and indeed extended its scope. Brown maintains that the "extroversion of the death instinct" involves "the drive to master nature as well as the drive to master man."[23] Whatever our attitude toward the theory of the death instinct may be, I think that Brown is correct in claiming that mastery over nature and mastery over men are closely related on the level of an ongoing psychological dynamic. Bacon's career in government service is not the source of that proclivity for political metaphors which determined his formulation of the idea of conquering nature; most of his predecessors and successors have found it equally congenial. The basic problem is not that the metaphor is inappropriate, but rather that it is simply taken for granted and rarely becomes the subject of critical reflection.

The measure of Bacon's success is indicated by the fact that the metaphor of "conquering" nature by means of science and technology appears everywhere today and seems so perfectly opaque both to learned social commentators and to the general public. In actual fact the metaphor's content is shifting and elusive, as an examination of Bacon's protean thought reveals. Even when the metaphor is clothed in full regalia, as it is in his brilliant utopian essay entitled *New Atlantis,* its manifest social content is rarely found worthy of comment. The usual references to that famous sketch would lead one to believe that it is no more than an interesting anticipation of modern scientific research institutes. This emasculation of it serves to detract attention from the only place in Bacon's writings where the idea of mastering nature is represented concretely in terms of its relationship to social progress. Especially when read in conjunction with an equally famous visionary tract, Thomas More's *Utopia,* the *New Atlantis* breathes life into the literary metaphors which we have been discussing.

4. Two Utopian Patterns

The careers of these two English philosopher-statesmen ran parallel in many respects. Both Saint Thomas More and Lord Francis Bacon rose through the state bureaucracy under rather disagreeable monarchs to become lord chancellor, although in each case their tenure of office was brief and ended in disgrace. Both executed their duties with a fairness that astounded their contemporaries, yet neither was immune from the particular vice of his day—in More's case the persecution of religious heresy and in Bacon's, a consuming secular ambitiousness. To their credit safeguards against these failings are conscientiously provided for in their visions of a better society.

A little more than one hundred years separates *Utopia* from *New Atlantis:* More's book was written in 1515 and the first drafts of Bacon's unfinished work (posthumously published in 1627) date from the period 1622–24. Very few competitors in the luxurious fields of modern utopian literature can rival them in terms of their impact on social thought down to the present time. Yet, although neither has managed to eclipse the other totally, the later sketch has struck the imagination of recent generations far more forcefully, and this impression continues to reverberate in ever wider circles. In the vast outpouring of studies and reflections on the social impact of scientific and technological progress, for example, one rarely encounters a contribution which does not refer at least in passing to the *New Atlantis.*

A remarkable consistency exists in the attitude of successive ages toward this mere fragment of a tale: the concept of methodically organized scientific research and the idea of a necessary bond between government and research organizations stirred men soon after Bacon's death and—as any survey of recent writing will reveal—elicits an equally fervent response today.[24] But those who pay homage to the inspiration of *New Atlantis* almost never examine the full, concrete context in which Bacon presents his vision of scientific and technological progress; this neglect is unfortunate, for the relationship between social institutions and the scientific research organization in *New Atlantis* is a decisive aspect of its message. The rather startling Baconian imagery can best be appreciated by contrasting it with the analogous situation in More's *Utopia.*[25]

More wanted to show that men's needs could be satisfied

by virtue of a more rational social division of labor. The inhabitants of Utopia are assured of the material satisfaction of their wants on the basis of a six-hour working day. The work period is sometimes shortened further, for the magistrates "do not keep the citizens against their will at superfluous labor," but rather stipulate "that for all the citizens, as far as the public needs permit, as much time as possible should be withdrawn from the service of the body and devoted to the freedom and culture of the mind." [26] This does not mean, however, that More is idealizing the "minimum state" described in Plato's *Republic:* his text clearly shows that wellbeing for the Utopians signifies a high level of material satisfaction of human needs. Rather, it is only the dedication to individual self-development as the overriding social objective which is responsible for the limitations imposed on the production of goods. For example, their clothing is said to be attractive but at the same time immune from the whims of fashion.

More is not extolling the virtues of contented savages by way of contrast to the civilized decadence of his own society. The Utopians are not the "happy sheep" of Tahiti scorned by Kant. On the contrary, their sciences of logic, arithmetic, geometry, and astronomy are said to equal or surpass those of Europe, and they apply scientific knowledge in the improvement of the soil and the treatment of disease. The essential fluidity of social relationships permits no barrier between those devoted to the advancement of knowledge and the rest of the population. No special merit is won by those who occupy posts of learning, and no rigid demarcation between functions is permitted: the learned can be relieved of their positions and can be assigned workmen's tasks, and workmen can qualify for the ranks of the learned occupations through diligent study in their free time.

Bacon's bold fancy conceived a far different setting for the relationship between social institutions and knowledge—the latter regarded in a restricted sense as organized scientific and technological research. In his story the crew of a European ship who have accidentally encountered the unknown island of Bensalem are allowed to witness the visit of one of the directors of their isolated research institute (named Solomon's House or the College of the Six Days' Works) to the island's main city. The leader of the European crew describes what he saw:

> The next Morning [the messenger] came to me againe, ioyfull as it seemed, and said; There is word come to the Governour of the Citty, that one of the Fathers of Salomons House, will be here this day Seven-night: Wee have seene none of them this Dozen Yeares. His comming is in State; But the Cause of his comming is secret. I will provide you, and your Fellowes, of a good Standing, to see his Entry. I thanked him, and told him; I was most glad of the Newes.
>
> The Day being come he made his Entry. He was a Man of middle Stature, and Age, comely of Person, and had an Aspect as if he pittied Men. He was cloathed in a Roabe of fine black Cloath, with wide Sleeves and a Cape. His under Garment was of excellent white Linnen, downe to the Foote, girt with a Girdle of the same; and a Sindon or Tippett of the same about his Neck. He had Gloves, that were curious, and sett with Stone; and Shoes of Peach-coloured Velvet. His Neck was bare to the Shoulders. His Hatt was like a Helmett, or Spanish Montera; And his Locks curled below it decently: They were of Colour browne. His Beard was cutt round, and of the same Colour with his Haire, somewhat lighter. He was carried in a rich Chariott, without Wheeles, Litter-wise; With two Horses at either end, richly trapped in blew Velvett Embroydered; and two Footmen on each side in the like Attire. The Chariott was all of Cedar, gilt, and adorned with Crystall; Save that the Fore-end had Pannells of Sapphires, set in Borders of Gold; And the Hinder-end the like of Emerauds of the Peru Colour. Ther was also a Sunn of Gold, Radiant, upon the Topp, in the Midst; And on the Topp before, a small Cherub of Gold, with Wings displayed.

The Chariott was covered with Cloath of Gold tissued upon Blew. He had before him fifty Attendants, young Men all, in white Satten loose Coates to the Mid Legg; and Stockins of white Silk; And Shoes of blew Velvett; and Hatts of blew Velvett; with fine plumes of diverse Colours, sett round like Hat-bands. Next before the Chariott, went two Men, bare headed, in Linnen Garments downe to the Foote, girt, and Shoes of blew Velvett; Who carried, the one a Crosier, the other a Pastorall Staffe like a Sheephooke: Neither of them of Mettal, but the Crosier of Balme-wood, the Pastorall Staffe of Cedar. Horse-Men he had none, neither before, nor behinde his Chariott: As it seemed to avoyd all Tumult and Trouble. Behinde his Chariott, went all the Officers and Principalls of the Companies of the Citty. He sate alone, upon Cushions of a kinde of excellent Plush, blew; And under his Foote curious Carpetts of Silk of diverse Colours, like the Persian, but farr finer. He held up his bare Hand, as he went, as blessing the People, but in Silence. The Street was wonderfully well kept; So that ther was never any Army had their Men stand in better Battell-Array, then the People stood. The Windowes likewise were not crouded, but every one stood in them, as if they had been placed.[27]

This portrait owes much to the attachment to pomp and ceremony which characterized Bacon's own life, but there is more in it than a mere autobiographical quirk. The narrative of the story quickly turns to a description of the affairs of Solomon's House, the more interesting aspect of the work to the majority of Bacon's readers, and it is easy to overlook the fact that in a few brief sentences the author has revealed a great deal about the general social arrangements in this island paradise.

In the pages of *New Atlantis* the structure of the relationship between scientific knowledge and its social context suddenly becomes problematic: not abstractly, but in fully concrete form as the problem of the interaction (or lack of it) between the director of the scientific research establishment and the general population. Although his luxurious attire and entourage clearly mark him as an extraordinary person, his lofty and severe countenance—he "had an Aspect as if he pittied Men"—and the silent blessing which he bestows on the watching multitude is even more indicative of the gap which separates him from his fellow citizens. This impression is reinforced by the news that the reason for his visit is unknown and that until this moment none of the directors had even been seen in the city for twelve years. The theocratic majesty of the visiting personage seems to stun the onlookers into an attitude of reverence and utter docility.

In an interesting article entitled "Bacon's Man of Science," M. E. Prior has offered a different interpretation of these passages. Prior argues that the portrayal of the leader from Solomon's House shows Bacon's recognition of the necessity for the scientist to be somewhat independent of the prevailing social interests, and he notes also the fact that the scientists of Bensalem are permitted to travel abroad in connection with their researches, whereas this is forbidden to the rest of the citizenry. The scientist represents the ideal of a "new man," the possessor of ethical standards far superior to those normally in authority, for whom "Bacon envisioned a new role in society and a society vastly improved by his dominant role in it." [28] This expectation rests on the assumption that the qualities essential for scientific research—impartiality, disinterestedness, analytical rigor, and so forth—would be operative also in the scientists' social role.

This argument seems to me a bit anachronistic in that it reads a modern perspective into Bacon's thought. Despite Prior's attempt to minimize the importance of religion, I think that Bacon did see religion, not the virtues engendered by scientific work itself, as constituting the primary source of ethical restraint on the uses to which knowledge might be put. Moreover, there is far too much psychological naïveté in the view—whether held by Bacon or by anyone else—that the ordinary passions of pride, self-interest, and ambition are absent from the scientific endeavor. The case of Newton offers perhaps an extreme example, but it is amusing to contemplate the ideal of a "new man" in the light of that great scientist's ruthlessly authoritarian exercise of bureaucratic power. No individual ethical qualities, be they ever so pure, can compensate for the lack of responsibility that is inherent in Bacon's scenario, where by comparison with the ordinary multitude the director of Solomon's House appears like a creature of a different species.

It is as easy to make too much of these passages as it is to slight this entire aspect of the story. There is no point in reading our contemporary impressions into a discussion of Bacon's intentions, for it is clear that he was not much interested in detailing the social and political arrangements in Bensalem. In the preface to the first edition, Bacon's literary executor, William Rawley, remarked that the author had intended to complete his sketch with a general description of the laws representing "the best State or Mould of a Commonwealth," and for this reason Rawley subtitled *New Atlantis* "a Worke unfinished"; but he also stated that in his last days Bacon had preferred to this task the furtherance of his chief concern, his natural history or resumé of scientific knowledge. Bacon's readers have generally upheld him on this point throughout the succeeding centuries, for the problem identified above has elicited very little comment: with a surfeit of speculation on the political frameworks of ideal commonwealths, they have been justifiably content to concentrate on the brilliant suggestion reflected in the portrayal of Solomon's House.

And yet it may now be time to break with this tradition. We must indeed acknowledge the influence exerted by Bacon's thought on the development of the idea of organized scientific research, as representing a permanent legacy which the industrially advanced nations have realized to a degree far surpassing the limits of the seventeenth-century imagination. As a theoretical and practical problem, however, the suppressed social dimension of Bacon's vision assumes greater significance from the perspective of the present. In Bensalem "the spirit of movement or of progress which dominates the intellectual life of the nation, appears to be totally absent from its political and moral life." [29] This fact is far more important for us today than it was for earlier periods.

The other outstanding utopian theorists of this period, Tommaso Campanella (*City of the Sun,* 1620) and Johann Andreae (*Christianopolis,* 1610), also emphasize the social importance of scientific and technological advances, and to some extent this separates all three from the spirit of More. In political terms, the scientist is subordinated to the established forms of authority, which in these ideal societies are still strongly religious in nature. In the description of Bensalem there is no indication that the social authority of the "Father of Solomon's House" overawes that of the city's governor whom he is visiting. However, in *New Atlantis* there is a suggestion that the scientific research establishment exercises complete control over its own activities and maintains an element of independence vis-à-vis the rest of society:

> And this we doe also: We have Consultations, which of the Inventions and Experiences, which wee have discovered, shall be Published, and which not: And take all an Oath of Secrecy, for the Concealing of those Which wee thinke fitt to keepe Secrett: Though some of these we doe reveale sometimes to the State, and some not. [30]

Again, this is no more than a passing remark and there is no direct evidence that it has made a special impression on later thinkers. Nevertheless it is crucial for any attempt to understand the social context of the mastery of nature, for clearly Bacon thought that those who were directly responsible for scientific and technological progress might also in large measure control its social utilization. Bacon shared a belief, common in the seventeenth and eighteenth centuries, that the individuals who championed the new scientific philosophy would necessarily also possess an ethical sense of responsibility for their work and a dedication to the task of developing more enlightened social institutions. His interpretation of the myth of Daedalus shows his acute awareness of the dangers inherent in the evil application of technical innovations and encourages us to believe that the secret deliberations within Solomon's House would be aimed at minimizing the possibilities for the misapplication of its discoveries.

One must emphasize that at the dawn of the "scientific revolution" this was by no means an unreasonable assumption. If there was a pervasive expectation that scientific knowledge and its technological applications would offer men the "power" to alter fundamentally the conditions of their lives, this hope was interwoven with the conviction that the power thus derived was essentially different in nature from that other, more familiar type of power—namely, the universal social and political domination of the many by the few. Bacon distinguished three kinds of human ambition: one which consists in advancing an individual's own concerns, another which seeks to promote the national interest, and a third "whose endeavour is to restore and exalt the power and dominion of man himself, of the human race, over the universe." [31] Scientific progress was to be the instrument of a species-ambition that subordinated and transcended the inclinations of lesser worth.

The comparison of More's and Bacon's utopias assumes renewed significance in the light of the historical fate of this expectation. In *Utopia* individual self-development pursued during the time not devoted to necessary labor—a time continually made more ample by the applications of scientific knowledge—is the primary objective: the "moral progress" of individuals was to be the mediating link between scientific progress and social progress, the third term in which is manifest the rationality of the whole. And it is just this element of moral progress that is so conspicuously absent in *New Atlantis*.

The problem does not consist in the lack of democratic control over the activities of Solomon's House, at least not insofar as the comparison with More is at all relevant. Neither utopia is ruled democratically. What is at stake is the nature of the interaction between scientific progress and social progress, the latter encompassing growth in physical well-being and in the capacity of individuals existing within the institutional framework of socialism to enjoy freedom, to accept responsibility for the common good, and to benefit from the absence of external compulsions. Thomas More offers us a glimpse of this possibility—admittedly it is no more than that —whereas Bacon does not. This is only a theoretical problem for the age in which these utopias were conceived, but it gradually becomes a pressing practical issue in the development of modern society.

From the perspective of the present we can see that it is not just the prominence assigned to scientific and technological activity that separates Bacon from More. Just as striking is the *fetishism* of that particular activity in *New Atlantis* as opposed to *Utopia:* this prognostication is indeed as brilliant as the more famous anticipation of scientific research institutes. The scene in which the exalted scientific administrator passes in solemn pomp before the awestruck populace can be read as an allegorical representation of our contemporary situation in which citizens are the passive beneficiaries of a technological providence whose operations they neither understand nor control. For all we can guess the citizens of Bensalem may be supremely grateful for the "increase of commodities" which results from the discoveries made in Solomon's House. But the isolation of that busy place symbolizes not only its physical remoteness but even more its complete "spiritual" separation from the concrete social life of the island's inhabitants.

No outstanding thinker after Bacon devoted comparable attention to the concept of mastery over nature. In the ensuing period it undergoes little further development, even though it is more and more frequently employed. The measure of Bacon's achievement is that most of those who followed him have found the form in which he cast this concept to be sufficient for their own purposes. So definitive was his work that the history of all subsequent stages in the career of this idea down to the present can be arranged as a set of variations on a Baconian theme.

4

THE SEVENTEENTH CENTURY AND AFTER

Nature becomes . . . purely an object for men, something merely useful, and is no longer recognized as a power working for itself. The theoretical cognition of its autonomous laws appears only as the cunning by which men subject nature to the requirements of their needs, either as an item of consumption or as a means of production.

MARX, *Grundrisse*

1. The Ideology of Nature

A fascination with nature marks the intellectual life of seventeenth-century Europe.[1] In popular tracts and learned tomes nature's praises are sung; the greatest scientists and philosophers vie with *littérateurs* and outright charlatans in estimating the prodigies of which nature is capable. The age is virtually obsessed with the notion that nature possesses "secrets" of inestimable value, and men insist that new methods of thought are necessary so that the "hunt" for them may be pressed into nature's hitherto unexplored lairs. Marvels and miracles were said to be locked up there, of such magnitude that once in possession of them men could imitate the

73

operations of the Creator.

The religious teaching that man completes and perfects the work of creation was reinterpreted along more "activist" lines. Nature was said to require the superintendence of man in order to function well, and this was understood as meaning a thorough transformation of the natural environment, rather than mere occupation or nomadic passage. This idea was used to justify the conquest and resettlement of so-called backward areas, such as the New World of the Americas, where it was claimed the native populations were not improving sufficiently the regime of nature. The belief that an active relationship between man and nature was one of the crucial elements in human civilization steadily gained momentum. It was a favorite theme throughout the eighteenth century as well, a common background for the various strains of Enlightenment thought: "Optimists and pessimists among the philosophes debated just how ready nature was to be dominated, but they agreed that whether man's relation to nature must be viewed as a collaboration or as a duel, that relation was intimate, inescapable, and exclusive." [2]

The glorification of nature was the ideological reflex of a new conviction, growing ever more firm among the intellectual circles that included the outstanding minds of the age, that the expansion of human power in the world was an objective worthy of their efforts; in other words, the belief that nature was capable of performing miracles, as yet unknown but accessible to human understanding through sustained research, and that the discovery of these hidden wonders would raise man's stature and dignity in accordance with the wishes of his Creator, was a powerful motivational force in winning adherents to the cause of scientific work. It served as a spiritual midwife in the great transitional period when what would later be called modern science (but then still known as "natural philosophy") slowly separated itself from magic, alchemy, and astrology.

The example of Descartes well illustrates how complex this process was. In later periods he came to be regarded as the founder of modern philosophical rationalism, and his invention of analytical geometry was recognized as an indispensable step for science; but to many of his contemporaries he had the aura of a master magician, and his praises were sung in astrological literature and similar sources. He himself advertised his philosophical "method" as both marvelous and easily mastered, a description which did nothing to discourage the wild fantasies common among the legions of enthusiasts in his day. The English philosopher Henry More wrote to him: "All the masters of the secrets of nature who have ever existed or now exist seem simply dwarfs or pygmies when compared with your transcendent genius." [3] Descartes followed the practice of sixteenth-century mathematicians in only publishing part of the solution to well-known mathematical problems, keeping the rest to himself. This was the accepted way of avoiding the embarrassment of seeing someone else publish an identical solution with the claim of having discovered it earlier—a not uncommon trick in those days of intense personal rivalry among leading intellectual figures. Others such as Christian Huygens used the alternative device of concealing a mathematical solution in a cryptogram or anagram whose meaning would be announced after some time had elapsed without anyone else having offered the correct answer to the stated problem.

Their contemporaries Pierre Gassendi and Marin Mersenne waged a vigorous, lifelong battle in an effort to distinguish "true science" from the many occult varieties whose protagonists flooded the literary market with fervent claims of proven results. [4] At the same time they had to establish the theological orthodoxy of the science of mechanics: the new synthesis emerged from the diverse currents of scholasticism, mechanism, and Renaissance naturalism. Mersenne combatted astrology, divination, and magic, but his fascination with them is shown by the fact that his books are filled with reports of prodigies and arcana—associated by him with the new sciences of statics, hydraulics, and pneumatics—which still strongly resemble those found in the magical and alchemical tradition. He propagandized ceaselessly on behalf of the sciences, calling upon the general public as well as the savants to abandon the old pursuits that were barren of results and to apply the new-found methods which promised to change the face of the earth. Also he bestowed extravagant praise on Hobbes's *De Cive* (1642), which more timid souls like Descartes found highly dangerous, precisely because of its subordination of religious controversy to the authority of the state. The advancement of science required social peace, and this could be achieved only if the state rigorously suppressed the open warfare arising out of the religious disputes that had plagued Europe since the Reformation: in Lenoble's words, "Mersenne loved science, social tranquillity, and God."

Gradually the content of what I have called the "ideological reflex" was transformed. Science and the mechanical arts (later technology) replaced "nature" as the focal point of the expectations associated with the expanding knowledge and control of natural phenomena; in other words, attention was shifted from nature as the source of marvels and new powers to the human instruments whereby these natural forces were discovered, integrated, and made serviceable for man's purposes. This too was the outcome of a long historical process during which there occurred dramatic changes in the philosophy of culture that ruled European civilization. Its roots are also to be found in the Renaissance. The great artists of that period, above all Leonardo, were responsible for a new, positive evaluation of the mechanical arts and of the technical skills of craftsmen which "made possible that collaboration between scientists and technicians and that copenetration of technology and science which was at the root of the great scientific revolution of the seventeenth century." [5] The drastic separation of theoretical and practical activity, which from its point of origin in Greek philosophy had permeated all of Western culture until that time, was only overcome after centuries of intellectual debate.

In the ongoing "quarrel of ancients and moderns," those who argued the superiority of recent times against the record of antiquity increasingly relied on the innovations of craftsmen and technicians as a conclusive demonstration of historical progress. These had long been regarded as "base" activities in contrast to the refined pursuits of high culture, but this attitude was subjected to sustained attack and eventually was reduced to a minority opinion. By the second half of the eighteenth century only hopeless reactionaries like the Jesuits were left to complain that the *Encyclopedia* was clearly the work of uncultured minds because it devoted so much space to the arts and crafts. Philosophers had begun to reflect on the methodological principles of the technicians and to insist that logic should incorporate them. "In these centuries there was continuous discussion, with an insistence that bordered on monotony, about a logic of invention conceived as a *venatio,* a hunt," Rossi writes; and, as opposed to the scholastic doctrines, this logic of invention "seemed above all concerned with projecting new methods and with extending the possibility of man's dominion over other men and over nature." [6]

By the beginning of the eighteenth century the moderns were gaining the upper hand. Their victory owed much to outstanding talents like those of Bernard de Fontenelle (1657–1757), whose long career was dedicated to vindicating for science and mathematics their right to be ranked among the principal contributors to culture and civilization. He maintained that science possessed certain values, notably skepticism, relativism, and pragmatism, whose cultivation would have a civilizing effect on the customary practices of human society. Its methodology tends to free the mind from the thrall of ignorance in general, he maintained, and if it were to be widely applied to the study of social relations considerable social progress would result. The "geometric spirit," he suggested in his famous "Preface on the utility of mathematics and physics" (1702), had suffused over the entire intellectual realm, bearing the virtues of order, neatness, and precision, and it had even lent aesthetic grace to works in the areas of morals, politics, and criticism. He also fought an obscurantist doctrine that rested on the age-old proclivity for using "nature" as the source of moral and aesthetic norms: "Often in order to scorn natural science, one throws oneself into admiration for nature, which is assumed to be absolutely incomprehensible. Nature is never so admirable nor so admired, however, than when she is known." [7] Fontenelle assured his contemporaries that the pursuit of science would not cause them to lose the aura of civility and refined taste which they cherished.

During the mature phase of the Enlightenment this message was transformed into a crusading ideology. The general surveys of that period, such as those by Ernst Cassirer, Paul Hazard, and Peter Gay, have documented the intense enthusiasm for science that gripped the majority of outstanding Enlightenment thinkers as well as the less able propagandists and epigoni who followed their example. The cause of human happiness was claimed to be identical with that of science, an opinion that did not seem to be qualified overmuch even within the strong undercurrents of pessimism found in Enlightenment thought. Men like Condillac, d'Alembert, and Condorcet brought to completion the idea that a single method could be applied in all the sciences, a method that was as valid for the study of society as it was for the investigation of nature.[8] A "social mathematics" or "geometry of politics," utilizing the mathematics of probability, was envisaged as a way of rationally reconstructing the system of social relations, since by these means human actions could be analyzed, laws of behavior formulated, and institutional arrangements leading to more humane conduct established.

The proposals for social reform based upon the new scientific methodology were engulfed in the political reaction following the French Revolution. But the ideology of science, shorn of social radicalism, survived and prospered in a theme which would recur frequently thereafter in European thought: the union of science and industry would by itself insure social progress and revolutionize the customary principles of social behavior. The theme was succinctly phrased by Baron Georges Cuvier, an outstanding scientist and secretary of the Royal Institute of France, in his "Reflections on the present state of the sciences and on their relations to society" (1816). Cuvier referred to the "universal opposition of rich and poor," to the jealousy among individuals which caused domestic unrest and the jealousy among nations that resulted in wars, and then declared that "industry and the science which produces it are the natural mediators" among these conflicts: "They equalize the nations in overcoming climatic obstacles; they draw together men's fortunes in making enjoyments more easily attainable"; and so forth.[9] This ideology took firm root and echoes still today.

2. Science and Industry

In seventeenth-century philosophy the concept of mastery over nature had achieved its definitive modern form, the one which has remained authoritative and substantially unaltered down to the present day. The rough path marked out by Bacon quickly became a well-traveled road. An age which was becoming enthralled with the prospects of scientific discovery found its guiding credo in the notion that man's dominion over the earth would be established by the progress of the arts and sciences. The new natural philosophy's objective was to insure "that *Nature* being *known,* it may be *master'd, managed,* and *used* in the Services of humane Life," wrote Joseph Glanvill in his *Plus Ultra* (1688), a defense of the recently founded Royal Society.[10] And despite the pervasive allure of magic and alchemy, with their seductive claim that nature's deepest secrets could be penetrated with a few clever strokes, a conviction was gradually taking root that the understanding of nature was won through the slow, painstaking drudgery of precise observations and controlled laboratory experiments. As exact science in the modern sense evolved, mastery of nature was more and more closely identified with it, until finally the latter ceased to have any meaning beyond what was attributed to science and technology as such.

This identification probably would have occurred even without the intervention of Bacon's writings, as the powerful currents of Renaissance philosophy and alchemy were assimilated during the formative years of modern science. But in fact his passionate espousal of the idea appeared just at that decisive moment, in the first half of the seventeenth century, when the foundations of the new science were being prepared by others. Thus the concept of mastery over nature was endowed with the extraordinary impetus which still sustains it. The price that was exacted for this service, however, was the surrender of the rich network of associations by which it had been supported previously. Only a thorough suppression of these earlier associations allowed this concept to survive the methodical purge of metaphysical residues which began in the seventeenth century, and to retain a place within the increasingly austere philosophical regimen prescribed by the modern scientific method.

The idea that the conquest of nature is realized through science and technology appeared more and more self-evident after the seventeenth century, and therefore few thinkers have felt the necessity of analyzing "mastery of nature" as a separate phenomenon. The meaning of that phrase has ossified, so to speak, by virtue of endless repetition within a widely accepted context, and these bonds must be loosened so that its own unique dialectic may be uncovered. To date this problem has gone largely unrecognized, except in the pioneering work of Max Horkheimer, founder of the "Frankfurt School" of social theory, which will be discussed later.

The customary formula in which the idea was encased is already evident in the matter-of-fact tone that characterizes Descartes's use of it in this very famous passage from his *Discourse on Method* (1637):

> But so soon as I had acquired some general notions concerning Physics . . . I believed that I could not keep them concealed without greatly sinning against the law which obliges us to procure, as much as in us lies, the general good of all mankind. For they caused me to see that it is possible to attain knowledge which is very useful in life, and that, instead of that speculative philosophy which is taught in the Schools, we may find a practical philosophy by means of which, knowing the force and the action of fire, water, air, the stars, heavens and all other bodies that environ us, as distinctly as we know the different crafts of our artisans, we can in the same way employ them in all those uses to which they are adapted, and thus render ourselves the masters and possessors of nature.[11]

For Descartes the new science is inherently linked with a practical philosophy, and in the sentence immediately following he declares that together they will lead to "the invention of an infinity of arts and crafts which enable us to enjoy without any trouble the fruits of the earth and all the good things which are to be found there."

This remarkable passage is often quoted. Unfortunately, there is no other one like it anywhere in Descartes's writings, and thus it is impossible to decide how much importance should be attached to the precise wording of his thought here. Of course this fact has not prevented many commentators from using a single quotation as a guiding thread in their interpretations of Cartesian philosophy. Writers in the Marxist tradition have expanded an incidental remark buried in the footnotes to *Capital* into elaborate arguments. Quoting most of the excerpt given above, Marx had commented: "Descartes's *Discourse on Method* shows that, just like Bacon, he saw a changed form of production and practical mastery of nature by man resulting from the changed methods of thought . . ." [12] Even thoroughly non-Marxist scholars have sought to make this passage the touchstone of Cartesianism, arguing that Descartes's reformation of thought really undertakes "the transformation of philosophy into the project of the mastery of nature." [13] Whether or not this contributes to the understanding of his thought is a matter which does not concern us here, but one thing emerges clearly from these interpretations, namely, that we do not learn anything further about the meaning of mastery over nature. Like so many others, Descartes simply does not devote sufficient attention to this idea to enable us to unravel its implications.

The same may be said of Saint-Simon's disciples, who flourished in the first half of the nineteenth century and who were the most enthusiastic propagators of an enlarged, updated Baconian vision. Around 1830 confident predictions of the attractive future being prepared for the human race on the basis of industrialization were already common. The Saint-Simonians went further and announced that the exploitation of external nature (*la nature extérieure*) under the conditions of modern industry and technology would radically alter the course of human history: "The exploitation of man by man has come to its end. . . . The exploitation of the globe, of external nature, becomes henceforth the sole end of man's physical activity . . ." [14] The characteristics of human activity in general were expected to undergo a change from warfare and competition to cooperation and peace. This revolutionary reversal, ending once for all the exploitation of human labor, was to be the direct outcome of accomplishments in industry, since the industrial form of production is according to them by nature peaceful and conducive to social harmony.

At that time their judgment was not quite as naïve as it appears now. But only ten years later Marx and Engels began their more penetrating studies on the structure and development of capitalism and industrialization. The concept of nature is one of the most important categories in all stages of Marx's work.[15] The interaction between man and nature through labor was for Marx the key to the understanding of history; the natural science and industry of the nineteenth century represented the most highly developed form to date of the ongoing "theoretical and practical relation of men to nature." The difficult task which Marx posed for himself was to show that this relation had a twofold aspect whose individual sides, so deeply interrelated, had to be distinguished.

On the one side, man is himself a natural being, and his capacity for labor is only a form of nature's energy; on the other, man seeks to transform nature so that his growing needs may be satisfied:

> He opposes himself to Nature as one of her own forces, setting in motion arms and legs, head and hands, the natural forces of his body, in order to appropriate Nature's productions in a form adapted to his own wants. By thus acting on the external world and changing it, he at the same time changes his own nature. He develops his slumbering powers and compels them to act in obedience to his sway.[16]

Marx presents here in outline the *dialectic* of man and nature. Nature is the "field of employment" for all human activity, the universal ground of the labor-process that is common to every form of social organization. In his activity man changes the natural world, but also is himself changed; his creative abilities unfold, opening up new possibilities for utilizing nature's resources, and the process continues indefinitely.

Marx envisaged a qualitative change in human development on the basis of the potentialities revealed by the industrial system already in the mid-nineteenth century. The replacement of labor-power by machinery would gradually free the individual from unending toil and allow the emergence of a new type of man who

> stands outside of the process of production instead of being the principal agent in the process of production. . . . In this transformation, the great pillar of production and wealth is no longer the immediate labor performed by man himself, nor his labor time, but the appropriation of his own universal productivity, i.e., his knowledge and his mastery of nature through his societal existence—in one word: the development of the societal individual.[17]

This dimension of Marx's theory, if taken by itself, could serve as an elaboration of the Saint-Simonian outlook, but since it offers only an account of the relation between man and nature, it is "abstract." What has been abstracted from is the dimension of class conflict which characterizes all of the more developed stages of the labor-process. The interaction of man with nature just described does not proceed spontaneously, as it were, but rather in response to compulsions arising out of antagonisms among the interests of men; moreover, at the higher levels of productivity the expression and satisfaction of needs is increasingly mediated by societal factors as opposed to purely immediate impulses. "To the extent that the labor-process is solely a process between man and nature," Marx writes, "its simple elements remain common to all social forms of development"; but in class-divided societies there emerge ultimately irreconcilable conflicts between "the material development of production and its social form" which finally result in the establishment of new institutions that alter the specific character of the labor-process.[18] Thus under capitalism men struggle with nature in order to satisfy their needs, but they do so in a prescribed way (namely, under conditions of wage-labor) that differs profoundly from other modes of production.

Considered abstractly, the level of domination over nature in any period is the same for all men, that is, it represents a stage of development attained by the human race as such. In reality, of course, the material benefits derived from the mastery of nature have always been distributed unjustly; but equally important is the fact that, however accomplished this human mastery becomes, the antagonisms of a class-divided society make it impossible for men to bring their productive system (of which mastery over nature is a part) under their control. This possibility emerges for the first time in a classless society. The realization of freedom, wrote Marx in the concluding section of the third volume of *Capital,* consists in "socialized man, the associated producers, rationally regulating their material interchange with nature and bringing it under their common control, instead of allowing it to rule them as a blind force." And Engels added that under socialism men will become for the first time "true masters of nature, because and insofar as they become masters of their own process of socialization." [19]

These statements are the concluding points in Marx's and Engels's overall social analysis, and together they represent the most profound insight into the complex issues surrounding the idea of mastery over nature to be found anywhere in nineteenth-century social thought or *a fortiori* in the contributions of earlier periods. They are inadequate as a starting point for contemporary discussion only because subsequent historical development has not yet taken the decisive turn forecast by them. To say this is not to disregard the substantial achievements of the socialist countries; but Marx and Engels could not have anticipated the degree to which a globally unified social order has become the only possible framework for bringing the material interchange between man and nature under rational control. They could not have foreseen that scientific and technological innovation would become a decisive instrument in the bitter struggle between capitalist and socialist nations, or that the "process of socialization" within socialism itself would be distorted in response to powerful military and ideological pressures emanating from capitalist societies.

Marx understood mastery of nature as a factor in the evolution of the labor-process. At an advanced stage of development this mastery is expressed in the fruitful marriage of science and industry. This approach lent great strength and cohesiveness to Marx's theory, for he was able to link different forms of the relation between man and nature with a theory of social change. In his own time he was justified in not presenting mastery of nature as an important individual social variable because (1) he expected that the general social consciousness of the proletariat would develop simultaneously with the domination of nature as a result of its labor-experience in industrial production, and (2) technology was not yet the source of false consciousness—a vital means of masking continuing injustice and class antagonism—within capitalist society that it was to become later. The state of both technology and class consciousness has changed decisively since Marx wrote, and thus his theory has had to undergo additions and modifications, one of the most essential of which is the reevaluation of mastery over nature.

3. Philosophy, Science, and Economy

More and more consistently, mastery of nature was being interpreted as the augmentation of human power in the world, and in concrete terms this process was thought to be clearly represented by the productive interaction of science and industry. The growth of science, of mastery over nature, and of human power thus have been regarded as identical phenomena. The contemporary philosopher-scientist C. F. von Weizsäcker has summarized this viewpoint in concise fashion: "The thinking of our science proves itself only in action, in the successful experiment. To experiment means to exert power upon nature. The possession of power is then the ultimate proof of the correctness of scientific thought." In another book he has explicitly called upon the standard Baconian formula in order to identify the "guiding principle" of modern science in its development since the seventeenth century: "It has become clearer than ever before that knowledge and power belong together . . . Pragmatic science has the view of nature that is fitting for a technical age." [20]

From this standpoint the difference between pre-modern and modern science appears as the outcome of both a changed methodology and a decisive alteration in the prevailing attitude toward nature. John Herman Randall expresses a fairly common judgment when he contrasts the approach of Aristotelianism, which dominated the natural philosophy of ancient and medieval times and which focused on understanding the "why" of things, with that of the modern period, which primarily asks "how" in its search for the means of transforming the world. Another nuclear scientist who has interested himself in philosophical matters, Werner Heisenberg, likewise suggests this contrast:

> At the same time the human attitude toward nature changed from a contemplative one to the pragmatic one. One was not so much interested in nature as it is; one rather asked what one could do with it. Therefore, natural science turned into technical science; every advancement of knowledge was connected with the question as to what practical use could be derived from it. [21]

Heisenberg thinks that this operationalist outlook has been completed in twentieth-century physics. He understands quantum mechanics and the uncertainty principle as signifying that the actual object of scientific research "is no longer nature itself, but man's investigation of nature," and that what it presents is a picture not of nature itself but rather of our relationships with nature.[22] In other words, contemporary physics cannot study the behavior of a natural phenomenon as it is in itself, entirely apart from the human effort to comprehend it, but rather must find its subject matter in the active interplay between man and external nature.

What are the reasons why science enabled men during the past three centuries to expand human power and to achieve a mastery over nature so much greater than anything known previously? What difference between pre-modern and modern science made this possible? The only thinker who ever attempted to give a comprehensive answer to these questions was Max Scheler, whose theory will be discussed in the next chapter. On the whole this problem has not been regarded as especially significant: for most writers the obvious facts of the matter—the demonstrated connection between scientific knowledge and power—have precluded any serious concern with possible explanations for it. The basis for this attitude had been established in the seventeenth century. The adherents of what was called the "new philosophy," who championed the principle of experimentalism in scientific research, demanded an end to the topics that dominated natural philosophy in the universities. Instead of debates on the "forms" and "essences" of natural phenomena, they proposed descriptive analyses leading to the formulation of laws of observable behavior. On the basis of these laws "directions for works" could be established, out of which would emerge mechanical devices rather than more volumes of empty speculations.

Gradually natural philosophy separated into two quite distinct pursuits, natural science on the one hand and the philosophy of nature on the other. The real work of science became identified with measurement, calculation, and mathematical demonstration, and the interpretation of nature "as such" was left to academic philosophers. The only philosophical principle that necessarily figured in the actual work of science could be stated in simple and straightforward terms:

What differentiates the nature of immediate sense experience from the nature of exact science, and what differentiates the scientific from the mythical concept of nature, is the characteristic principle of selection which operates within each. None of these conceptions simply copy the concrete totality of appearances; rather, each picks out certain elements of the totality as being decisive and essential. Each groups the totality of phenomena around specific conceptual centers and arranges them into different levels of relations.[23]

A comparative study of these "principles of selection" in relation to their implications for human activity, however, would be completely foreign to modern science "as such" and instead would fall in the domain of the philosophy of nature. Indeed it was the tradition of German *Naturphilosophie* that, in the entire course of its development from Herder and Goethe to the last period of Schelling, endeavored to fashion a conception of nature that could integrate the operative principles of the natural sciences with the other perspectives of human experience, such as ethics and aesthetics. The great Hegel scholar Johannes Hoffmeister interpreted the general tendency of *Naturphilosophie* as an attempt to transcend the attitude that nature existed only as an object of domination for man, an attitude that the philosophers of nature regarded as having become prevalent as a result of the enthusiasm generated by the natural sciences.[24]

Interest in the philosophy of nature declined rapidly after the middle of the nineteenth century, and the challenge to the idea of the domination of nature from that quarter has never been taken very seriously since then. Mastery of nature had wedded its fortunes to those of science, and every success of the latter was regarded *ipso facto* as a further validation of that idea. The philosophy of nature yielded to the philosophy of science and later to the sociology of science, an intellectual transition which reflected the growing social importance of sustained scientific and technological innovation. Both the origins and the principles of science had to be clarified so that its triumphant methodology, once cleansed of any remaining philosophical errors, could penetrate more easily the other areas of social endeavor.

Even within the special discipline known as the philosophy of science, the question concerning the philosophical changes which made possible the marriage of science and industry never became prominent. The only full-length study of it is one in the Marxist style done by Franz Borkenau. He took his cue from a passing remark by Marx (in the footnote to *Capital* referred to earlier) to the effect that Descartes, in conceiving animals as machines, was "seeing with the eyes of the manufacturing period." Borkenau contended that the decisive branch of the new science of the seventeenth century was mechanics, and that mechanics can best be understood as the science of the manufacturing period, that is, as the "scientific treatment of the manufacturing production process." Out of its basic concepts, as developed by Galileo, Descartes, and others, was formed a new world-view, and this "mechanistic picture of the world is an extension of the processes in manufacturing to the entire cosmos." [25] This relationship is according to him not an accidental one; rather, the manufacturing system is a necessary presupposition of the new mechanics.

His argument rests on the concepts of "abstract labor" and "abstract matter." By the former is meant the characteristic aspects of human labor under conditions of capitalism as described by Marx in the opening chapter of *Capital*. Manufacturing and capitalist commodity production gradually destroyed the hand-craft productive process, in which goods reflected the particular qualities of the individual craftsmen who fashioned them, and replaced it with a system of "abstract-general" labor, that is, one wherein all differentiation of skill, strength, training, and so forth tends to disappear. The category of labor time, a purely quantitative measurement representing an averaging of specific qualities dispersed throughout a large working force, is the basis upon which the labor costs of production are calculated. "Abstract matter," on the other hand, refers to the concept of matter developed by Galileo and Descartes as part of their mathematical physics. In order to formulate the laws governing the movement of natural objects, they suggested that it was necessary to disregard the sense-qualities of things (those apparent to sight, hearing, touch, taste, and smell) and to posit the existence of a uniform substance common to all objects. Since this substance or matter is assumed to be everywhere the same, the differences between things can be reduced to simple quantifiable proportions—an idea later expressed in the concept of "mass." As only relations of quantity are involved, the laws of motion could be set down entirely in mathematical and geometrical terms.[26]

Borkenau's main point is that the practical and technical aspect of this development, namely, the emergence of abstract labor in the manufacturing process, had to accompany the theoretical system, that is, the formulation of the concept of abstract matter in mechanistic philosophy and science. The two are inseparable; the theoretical breakthrough could not have happened of itself, with no relation to other forms of social activity. In his view "only the application of capitalist methods in the labor process makes possible the observation of nature according to quantitative methods." Actually Borkenau is not proposing a simplistic determinist account, for he explicitly denies that the beginning of the manufacturing process was the "cause" of the mechanistic world-view. Rather, both arose together and interacted constantly, the progress of each conditioning and encouraging that of the other. The principles of mechanistic physics were generalized in order to form the basis of a comprehensive metaphysical system, embracing "methical rules for the expansion of domination over nature" and thus opening up enormous possibilities for the extension of manufacturing into different areas.[27] In turn the latter, gradually conquering the entire system of economic relations and demonstrating daily the practical results of the new science, helped the mechanistic philosophy to overcome its philosophical and religious opponents.

Borkenau's book was sharply criticized in an article by Henryk Grossman, who disputed his identification of quantificational methods with the rise of manufacturing. Grossman tried to show that already in Leonardo's thought there is an emphasis on exact methodology, and this example makes evident the real source of the quantifying spirit—the machine. Beginning with Leonardo and culminating in Descartes, where the image of the machine becomes the shaping force of a metaphysical world-view, "mechanistic thought and the improvements of scientific mechanics during the 150 years of its development since the middle of the fifteenth century show nowhere the traces of a close relation to the breaking-up of labor in the manufacturing process, but on the contrary always and everywhere the strictest relation to the use of machines." [28] The manufacturing process multiplies the productivity of human labor by means of machinery, but in order to do so it requires the factory laborers to become virtual appendages of these mechanical devices, destroying their individuality in the course of augmenting an output which is measured basically in quantitative terms. Thus according to Grossman it is the machine apparatus which is the decisive factor in modern production and also in the formulation of the conceptual framework of theoretical mechanics.

Of course this debate offers only a sample of what is by now an extensive literature. A correct understanding of the link between modern science and social change has been a matter of primary concern for the official Marxism of the Soviet Union, for example. [29] In Europe and North America academic controversy has largely centered around a thesis— first hinted at by Max Weber and later argued by Robert Merton—associating the rise of science with Puritanism. [30]

And Joseph Needham has explored in a series of essays (collected in his *The Grand Titration: Science and Society in East and West*) the problem of why modern science developed in the West rather than in China.

The use of Leonardo in Grossman's critique demonstrates how problematical the search for historical origins may become, for one cannot assume that there is any necessary continuity between Leonardo's attitude toward machinery and that of a later period. In Paolo Rossi's opinion the papers and notebooks of Leonardo's which detail his work on machines reveal a larger context that clearly separates him from his successors in the seventeenth century. Not production and manufacture, but diversions for festivals and entertainments, are the objectives of Leonardo's devices, showing that "he was interested in machines more as the result and proof of human intelligence and genius than as a means for the actual mastery of nature." [31] By themselves neither capitalist industry nor the science of mechanics nor machinery are necessarily related to the idea of mastery over nature, and likewise this idea by itself cannot serve as an explanation for the development of those phenomena (either in isolation or in terms of their interrelationship) throughout modern history. During the past century and a half this idea indeed has become identified more and more closely with science and technology, but viewed retrospectively this association affords only a partial insight into the intellectual biography of earlier periods, when the hegemony of traditional religious and philosophical perspectives was still unshaken.

Alexandre Koyré was perhaps the most sympathetic contemporary critic of the "psychosociological" interpretations outlined above. He recognized a general tendency to explain modern science as the outcome of a larger transformation in social and intellectual life during which "activity" replaced "contemplation" as the highest value: "Modern man seeks the domination of nature, whereas medieval or ancient man attempted above all its contemplation." [32] Koyré agreed that, despite the impressive level of continuous technological achievements during the Middle Ages, there is a profound difference between its predominantly other-worldly outlook and the concern for the improvement of earthly life which comes to the fore in modern times. Similarly, it is undeniable that an interest in technics and practical problems is ever more firmly joined with theoretical science. But he believed that purely theoretical motivations were at work in at least two decisive areas, namely, mathematics and astronomy, and that it was a new metaphysics that guided the breakthrough of the scientific revolution in the seventeenth century. This metaphysics was characterized by the destruction of the ancient idea of the "cosmos," that is, the notion that the universe was arranged according to tightly ordered gradations of higher and lower spheres, and by the mathematization or geometrization of space.

For Koyré, Bacon was the "announcer" of the new science, although he was not one of its founders. As we have seen, the notion of the domination of nature was attached to the new science by Bacon and others and is often referred to in passing by some of the most important figures in the scientific revolution itself, such as Descartes. However, it cannot be considered a determining factor in the actual origins of that science or in its subsequent conceptual development. But it does provide a guide to the subjective motivations of the intellectual leaders during this time; and, to the extent that their expectations subsequently become widely accepted in European civilization as a whole, the conviction that the mastery of nature would effect a beneficent social transformation became a powerful ideology in modern society.

4. Science and Society

The major historical components of the idea of mastery

over nature have now been indicated. Although many additional references to it could be cited, in my opinion these would only provide further illustrations for the lines of development established here. Moreover, the history of this idea cannot be tailored neatly, since the sources on which it is based are themselves extremely diverse and resist precise categorization. Thus many individual expressions of it contain interesting nuances which elude the traps and filters of any schematic presentation, such as the one attempted in the preceding chapters. A really satisfactory biography of the idea would have to display these shadings fully; it has not yet been written. The fragmental biography offered here is intended specifically to supply the requisite background for an analysis of the contemporary usage and significance of the idea.

This analysis constitutes the subject matter of Part II. The connection between mastery of nature, science and technology, and social progress will be systematically explored in the light of three twentieth-century contributions: (1) Max Scheler's concept of science as *Herrschaftswissen,* "knowledge for the sake of domination"; (2) the distinction between nature as the subject of scientific research and nature as the background dimension of everyday life which is found in Husserl's phenomenology; (3) the radical social theory of Max Horkheimer and his co-workers. None of the three have so far been accorded a proper place in the sociology of science, which has become an increasingly important specialized discipline within the past fifty years, although they are most relevant to a critical analysis of the interaction between science and society.

Some time ago J. D. Bernal wrote: "The tasks which the scientists have undertaken—the understanding and control of nature and of man himself—is merely the conscious expression of the tasks of human society." [33] This was the concluding message in a book that Bernal felt compelled to publish in 1939 because in the aftermath of the First World War and the Great Depression science was being attacked as a basically harmful force in society. Such negative voices may be heard occasionally even today, but normally they are drowned in the happier choruses led by the celebrants of scientific progress. The appeal of Bacon's vision is undiminished. Recently Steven Dedijer recalled the Lord Chancellor's work in presenting his case that more attention must be paid to the "science of science," a specialized field which concentrates on identifying the cultural conditions responsible for the flourishing of science. Since the methods of scientific research "are today the basic tools used to extend man's knowledge and power, not only with respect to nature but also with respect to society," we need (in his view) to study the psychological complex of attitudes and values that encourages scientific work.[34] The "deliberate promotion" of the social and behavioral sciences is especially desirable, for according to Dedijer these are now recognized as one of the essential keys to the solution of those global social problems that become more pressing every year.

Dedijer examined a group of social research policy reports prepared by academic experts for governmental agencies in several different countries. All of them, he found, "assert that the major objective of the social sciences is to aid the understanding and control of the present rapid rate of social change"; in addition "all imply, and three state categorically, that this rapid social change is due mainly to rapid scientific change."[35] In these studies the conquest of nature through science and technology is seen as the primary cause of social instability at present, thus giving rise to the necessity of developing techniques for "controlling" societal mechanisms. Since the seventeenth century the concept of mastery of nature has always had this hidden dimension, permanently concealed from view like the dark side of the moon: mastery in society, that is, mastery of social change. Paradoxically, however, it is just the recognition of this hidden aspect that enables us to frame the basic questions to be pursued in a critical analysis of mastery of nature. To consider the problem of mastery in society we would ask: What kind of mastery? And what kind of society? The analogous questions for mastery of nature are: What kind of mastery or domination is achieved with respect to nature? And what is the "nature" that is mastered in this process?

PART TWO

Science, Technology, and The Domination of Nature

5

SCIENCE AND DOMINATION

> *Science—the transformation of nature into concepts for the purpose of mastering nature—belongs under the rubric "means." But the purpose and will of man must grow in the same way, the intention in regard to the whole.*
>
> NIETZSCHE, *The Will to Power*

1. The Concept of Herrschaftswissen

Any critical examination of the idea of mastery over nature must confront the thesis that has shaped the common understanding of this notion for several centuries: the conquest of nature by man is achieved by means of science and technology. Only by carefully appraising this thesis is it possible to show that the full dimensions of what is intended in the human mastery of nature have been obscured because of it, for upon examination the common understanding reveals a host of ambiguities and unclarified premises. Such an operation is necessary in order to clear the way for a discussion of the concrete issues which are implied in the following question: what are the *specific* human objectives that are sought in the domination of nature? The conventional answers—control over the environment, the augmentation of human power in the world, "relief of the inconveniences of man's estate," and so forth—will no longer suffice. Their intolerable vagueness

101

helps to conceal a set of fundamental social contradictions.

Looked at from a different perspective, this familiar thesis might also appear as a misrepresentation of the significance of the modern scientific and technological enterprise itself. In other words, if upon examination the notion of conquering nature through science and technology turns out to be largely vacuous, in other words, a delusive image, we may be prompted to search for different categories to define the meaning of this enterprise. Certainly the past should not be judged by contemporary standards alone, and I am not suggesting that this image has always been "wrong"; on the contrary, it was once a valuable imaginative force which mobilized support for a radically new approach to the investigation of nature. On the other hand, the liberating slogans of the past, by virtue of their very success, may become intellectual straitjackets under changed historical conditions. This is not the place to speculate on what the other categories referred to above might be; the analysis to follow focuses on the prior problem of discussing the qualifications which should be attached to the representation of science and technology as instruments designed for the conquest of nature.

To attack this problem it is necessary to deal with science and technology as separate phenomena, at least to some extent, even though they are invariably linked when they are associated with the mastery of nature. The reason for this procedure is that (as I shall try to show) the kind of mastery relevant to each is quite different, even potentially contradictory under certain circumstances. This separation is an analytical device, the value of which will be tested in the course of the exposition; it is a conscious "distortion" of the prevailing reality wherein science and technology daily display their productive interaction. It corresponds to the distortion that is implicit in the concept of mastery of nature itself: each particular historical period aims not at "mastery of nature" *per se,* but rather at realizing a number of determinate steps in scientific and technological capabilities; viewed retrospectively and prospectively, however, the whole process is described as the conquest of nature. Thus in a sense this phrase distorts the character of the reality which is immediately present, while simultaneously functioning as a way of calling attention to various underlying tendencies that may not be otherwise apparent.

In partially divorcing science from technology for the purposes of this discussion I do not mean to contradict the theory concerning the "internal" connection between the two which was first stated in modern form by Spengler, then elaborated in the 1920's by Max Scheler, and subsequently reformulated in various ways, most recently by Marcuse in *One-Dimensional Man*. According to this theory, modern science developed on the basis of inherently operational concepts which were suitable *a priori* for technological application, and this is the feature that distinguishes it sharply from preceding conceptions of science. Important philosophical issues have been raised in this thesis, but because they are not directly relevant to the present topic no special notice will be taken of them. The historical inseparability of modern science and technology is taken for granted throughout this study, and in fact it is just their inseparability that necessitates a precise statement of the different senses in which mastery of nature may be used with respect to each.

Those who regard the conquest of nature as a social goal of modern society, and who adhere to the widespread view that the natural sciences provide the means by which this objective has been pursued during the past few centuries, would seem to have in Max Scheler's sociology of knowledge an ideal presentation of the historical and philosophical bases for those ideas.[1] There is no discussion of this theme more directly relevant, more thorough, or more intelligent than Scheler's. The relatively brief arguments concerning the connection between science and the domination of nature offered by other writers usually turn out to be dependent on his more elaborate undertaking; on the other hand, the few attempts to expand his treatment along different lines have not added any significantly new dimensions to our understanding of this topic. Above all, what is primarily lacking in the works of the many authors who make passing reference to the conquest of nature is any attempt to describe concretely what they think is meant by this phrase; at the very least Scheler's argument enables us to come to grips with this elusive idea.

Surprisingly enough, the available evidence indicates that most of those who have been concerned with this topic are completely unfamiliar with Scheler's writings. This is regrettable, for there is much to learn even in the process of disagreeing with his exposition. Ludwig Landgrebe is a notable exception in this respect, having recognized the significance of Scheler's analysis of the domination of nature for the pervasive concern with values that characterizes recent European philosophy. Landgrebe remarks that the proposed solutions for the crisis of values usually are mere restatements of nineteenth-century viewpoints, and that in order to improve matters thinkers will have to recognize the connection of value-problems "with those of technology and the technological interpretation of the world." He claims that the second set of problems arose out of the profound influence which was exerted on all modern thought by the idea of human mastery in the world: "While at the beginning of modern times the definition of man as *animal rationale* was still generally accepted, his *ratio* [reason], deployed in science, was now understood as the power of gaining mastery of the world. The course of the evolution of the modern mind was thus interpreted as involving the self-realization of man by virtue of the power of his scientific reasoning." [2]

He has also summarized Heidegger's viewpoint, noting that the latter has acknowledged his indebtedness to Scheler in his writings on technology and the scientific-technological mastery of the world. Landgrebe claims that Heidegger deepened Scheler's analysis by relating the contemporary problem of technology to certain fundamental continuities in the entire history of Western metaphysics. But Heidegger's ideas on this subject have remained closely bound to Nietzsche's concept of the "will to power," and in my view they are not as important as the other sources used in the following pages for an understanding of mastery over nature and its relationship to scientific-technological progress.[3] On the other hand, although (as we shall see) Scheler too was obviously influenced by Nietzsche's concept, he applied it creatively in a philosophical investigation of modern science, which he regarded as an instrument designed for the domination of nature. Scheler's writings are still the most important single source for an analysis of the relationship between science and domination.

The part of Scheler's work which is of interest here is his elaboration of the concept of *Herrschaftswissen*, "knowledge for the sake of domination." Because Scheler himself does not give a systematic exposition of it, because it is a difficult concept which Scheler formed from wide reading in the history of philosophy and the philosophy of science, and finally because it is so unfamiliar to those who are interested in the theme of the mastery of nature, I have devoted a considerable amount of space to presenting an account of it before trying to assess its merits.

References to the notion of *Herrschaftswissen* can be found throughout many of Scheler's books, and it is among the principal themes in two of his long essays, "Problems of a Sociology of Knowledge" and "Cognition and Labor," collected in a volume entitled *Die Wissensformen und die Gesellschaft* (*Society and the Forms of Knowledge*). He never wrote either a book or an article that was specifically concerned with this concept, however, and what follows is an attempt at a synthetic construction of it based on references drawn from scattered points in these two essays.

2. Science and the Will to Power

In every historical period man has pursued the struggle with nature in order to maintain his existence. To do this he attempts to "dominate" the surrounding environment in the sense that he regards it from the point of view of its usefulness for his existence; this is obviously a necessary process and one which man shares with all the more highly developed forms of organic life. This historical constant in human history, represented in conscious form—the development of various techniques for subjecting the environment to the ends of man —Scheler refers to as *Herrschaftswissen* (or the "positive sciences"). It coexists with two other types of knowledge, namely, metaphysics and religious thought. These are also historical constants, that is, all three are present in some form in all stages of human civilization and in every developed culture. They do not follow each other as successive epochs in a scheme of linear historical progress, as Comte had argued.[4]

At the outset one must note the impact of Nietzsche's fragmentary remarks concerning the "will to power" on Scheler's concept of *Herrschaftswissen*. In one of these passages Nietzsche wrote:

Knowledge works as a tool of power. . . . In order for a particular species to maintain itself and increase its power, its conception of reality must comprehend enough of the calculable and constant for it to base a scheme of behavior on it. The utility of preservation . . . stands as the motive behind the development of the organs of knowledge—they develop in such a way that their observations suffice for our preservation. In other words: the measure of the desire for knowledge depends upon the measure to which the will to power grows in a species: a species grasps a certain amount of reality in order to become master of it, in order to press it into service.[5]

Nietzsche's basic intention was to show the primacy of valuation in all forms of human experience—in religion and aesthetics as well as in logic and metaphysics. The basic impulse is to impose order on the chaotic field of immediate sensation and perception, the world of things in process of becoming, so that calculation and prediction may be gradually extended in scope. The forms of reason enable us "to misunderstand reality in a shrewd manner," that is, to create a stable basis for experience and action in order to assure the preservation and enhancement of life.

Scheler accepted the idea that a positing of values constitutes the foundation of human thought, but he modified considerably Nietzsche's account of this process. In presenting a tripartite scheme for the forms of knowledge, only one of which (*Herrschaftswissen*) specifically reflects the drive to dominate the world, he weakened the force of Nietzsche's contention that all means of cognition are different manifestations of the will to power. On the other hand, what is mainly lacking in Nietzsche's conception is an indication of the various stages through which the will to power has evolved. For although the "biological" necessity for preservation of the species and the innate human need to enhance the conditions of life are always operative, these requirements may be met in any number of ways. Scheler tried to show that this is in fact the case. For most societies and civilizations up to the present century, the techniques of species-preservation have remained on a relatively crude and primitive level, and this is true of Western civilization throughout the medieval period: the actual transformation of the environment, when compared with the achievements of the last two centuries in this regard, was minimal. Accordingly, the positive sciences, whose basic objective consists in "the technical projection of possible action in the world," [6] have been traditionally subordinated to the other modes of knowledge. In the Aristotelian system, for example, the positive sciences are established as parts of an overall metaphysical world-view (the Aristotelian "first philosophy" is the metaphysical system *par excellence*); and similarly in other cultures characterized by a low level of mastery over the natural environment, intellectual life is governed by philosophical and religious systems which devalue the concerns of mundane life.

Scheler describes the Aristotelian outlook as "organological," one whose basic conceptual structure was rooted in organic life and its various modes of being, and he insists that an "apractical-contemplative" attitude governed the way in which Aristotelian thought attempted to grasp the structure of being. The world-view which supplanted it in European civilization beginning in the seventeenth century contradicted its predecessor on these points, and the question of what can explain this profound transformation is for Scheler a crucial one. A change in intellectual attitude of such far-reaching significance could indicate nothing less than a fundamental alteration in the life of man in Western civilization. Such a decisive alteration did occur, according to Scheler, in European history in the transition from medieval to modern society. It manifested itself simultaneously in the relations of social classes, the economy, political institutions, and in all the realms of intellectual life—religion, ethics, political thought, natural science, and so forth. Scheler concentrates on the changes in intellectual matters and relates them to a cursory sketch of social, economic, and political developments; among these changes he finds the one pertaining to the attitude toward nature (or the external world) to be decisive.

The major conceptual differences between the new scientific philosophy of the seventeenth century and the natural philosophy which it replaced can be expressed in a variety of ways. Scheler summarizes his conception of the essential difference as follows:

And in place of a search for a categorical hierarchy of things (the scholastics) and a classificatory pyramid of concepts yielding a teleological "realm of forms" there is a search for quantitatively-determined, lawfully-ordered relations of appearances; the prevailing conceptions of "type" and qualitative "form" give way to quantitatively-determined "laws of nature." [7]

Within the framework of this general transformation many particular shifts in ways of thinking took place. Among them are the primacy of the category of quantity over that of quality and of relation over that of substance and its accidents, the substitution of inertial motion governed by geometrical principles for Aristotelian local motion, the development of analytical geometry, and—in a sociological sense—the growing social authority of the scientific researcher over that of the cleric and the "learned" man who cultivated the liberal arts.

Scheler's evaluation of the seventeenth-century philosophical and scientific revolution can be gauged from what has been outlined above. This change represents the liberation of the positive sciences from the tutelage of metaphysics and the establishment of a foundation for a set of techniques capable of rendering nature more subservient to human purposes. The new science viewed itself as freeing the investigation of nature from all "metaphysical" and "religious" assumptions and dogmatism. The historical result of its efforts in this regard is a novel conception of science: scientific knowledge is a type of understanding which stands apart from all value-judgment and value-determination, and the objects of scientific knowledge are themselves necessarily value-free. In Scheler's view this is the key to understanding modern science as the highest possible development of *Herrschaftswissen:* "To conceive the world as value-free is a task which men set themselves on account of a value: the vital value of mastery and power over things." [8] The positive sciences, no matter how primitive they might be, are governed by a primary value in their approach to the world, namely, the attempt to increase the power of man over the environment. One of the great discoveries of modern thought was the paradoxical conclusion that this primary value could best be served by viewing the world as entirely value-free, that is, as consisting entirely of matter potentially transformable for purposes of human use.

For Scheler the intellectual changes which began in the seventeenth century were truly revolutionary. Taken together, they may be seen as constituting a different "principle of selection" or "principle of abstraction" vis-à-vis natural phenomena in comparison with the principles of the science which it supplanted. What is the actual difference between these two modes of abstraction? In Scheler's view, on one point a comparison is decisive: the new science has enormously expanded the ability of man to exert his mastery over the environment. One means of understanding the mode of abstraction which guides modern science is to realize that it devalues the cognitive significance of all those things (sense-qualities, final causes, aesthetic values) which do not aid in man's domination of things; simultaneously, it asserts the cognitive priority of those aspects of natural phenomena which fit the scheme of prediction and control, for

> . . . even the simplest sensation and perception of our natural perceptual world is already so shaped by the scientifically-fixed channels into which our impulses and awarenesses are directed that relative constants and temporal uniformities of the actual processes of nature have an unquestionably greater prospect and chance to become "registered" through sensations and perceptions than do relatively inconstant and temporally unique phenomena . . .[9]

Scheler claims in this passage that the positive sciences are characterized by a built-in prejudice in favor of relatively constant and uniform natural processes because those are the most useful in developing a set of techniques whereby one can predict the outcome of a plan with relative certainty and

thus choose the means by which the environment m
utilized in accordance with human needs and desires.

An *a priori* principle of selection is at work in the obse
tion of natural phenomena which provides the foundation
a new concept of science and objectivity. Scientific knowledge
is attained only in respect to observably measurable phenomena: a proposition makes no sense whose affirmation or
denial would not result in a difference that could ultimately
be expressed in some measurable form. Out of the totality
which is given in experience certain factors are isolated:
bodies, movement, spatiotemporal causality, and relations of
magnitude. These factors are then united and related in a
symbolized manner—for modern science, in the language of
advanced mathematics; this symbolical representation is constantly being developed and refined in order to devise an
increasingly accurate account of the behavior of these aspects
of the totality which have been admitted into the system by
virtue of the operating principle of selection.[10] Those aspects
of experience excluded by the selective mechanism, that is,
those aspects which cannot be represented in the chosen sym-
bolic language of mathematics (such as aesthetic intuition), are not merely considered as an additional order of
experience, but rather are assigned a fundamentally different
status: they belong to the realm of the "subjective" and the
"unscientific."

The final important point in Scheler's argument is the attempted demonstration of an inner connection between the
theoretical scientific structure and the technological applicability of science. Scheler contends that technology is not the
subsequent application in practice of a "theoretical-contem-
plative science"; rather, *Herrschaftswissen* is primarily characterized by the unity of its theoretical and practical aspects.
The actual forms of thought and intuition which make up its
conceptual apparatus themselves operate according to a principle of selection guided by the practical objective of asserting

mastery over the environment. This does not mean that the science is governed by specific or immediate technological goals which shape its theoretical structure, but only that in general it embodies the drive for power (*Machttrieb*). Modern science represents the highest possible development of this drive, for it strives to convert the entirety of nature into a field of operation for exclusively human purposes; it aims, in Scheler's words, "to construct all possible machines—to be sure at first only in thought and as a plan—through which man could lead and bind nature to whatever ends, be they useful or not, that he desired." [11]

There is thus an indissoluble link between human theoretical and practical activity: the elementary practical requirements of the struggle for existence shape the categories of thought, and the relative adequacy of the cognitive apparatus determines the range of success for human practice in the world. Scheler insists, however, that this does not mean that all scientific perspectives are equally arbitrary or subjective. The limits of the theoretical-practical activity of man are set by the ontological character of the world. If the world did not possess, as part of its ontological structure, a mechanistic aspect—in other words, an aspect which could be represented by uniform laws of the behavior of matter expressible in mathematical form—no drive of the human will to power could make it so. [12] The attainments of the modern natural sciences represent a conclusive proof of their mastery of a significant segment of reality and a clear demonstration of their superiority over all other techniques (for example, magic) for transforming nature in accordance with human purposes.

Although the basic objective of the positive sciences always remains the same (mastery of the environment), their condition varies greatly from one historical-cultural epoch to another. Scheler's chief interest is in the transition from the medieval to the modern period in Western history, and he claims to find in that transition a fundamental value-transformation which was eventually expressed in every aspect of material and intellectual life. The drive for power in the Middle Ages concentrated on the exercise of domination over men, whereas the new drive sought power over things, more precisely the means for transforming things into valuable goods. Domination of nature became the primary focus of the will to power in the modern period, subordinating the domination of men, and this change signifies above all the victory of a new ethos and a new structure of drives.[13] The new science of this period and the logical system of categories on which it is based have their original foundation in the ethos whose overriding interest is the domination of nature.

In asserting a link between the formation of the new science on the one hand and the structure of modern society on the other, Scheler did not think he was advocating a particularly novel or provocative thesis. He believed that the important works of Simmel, Tönnies, Sombart, Dilthey, and Bergson had established

> the necessary connection of the domination of a quantitative and predominantly mechanical view of nature and the soul (which takes all qualities from bodies) with the increasing domination of industry and technology . . . and the simultaneous connection between the same quantificating view of the world and the money-and-acquisition-economy in which goods, stripped of all qualities, become a "commodity." [14]

But there is no simple relation of causal determination between intellectual forms and social structure. Scheler speaks of the positive sciences as sociologically codetermined. He means that, whereas the basic structure of human cognition and its system of categories is not dependent upon any specific form of social life, the *selection* of a particular way of looking at the world and the valuation of that mode of thought above all others can only be understood sociologically: the formal-mechanical system of thought "is the product of pure logic (plus pure mathematics) and a pure valuation of power in the selection of the observable phenomena of nature. And only in this second power factor lies also the sociological codetermination of this selective principle of the appearance of nature." In a similar spirit Max Horkheimer remarked that it is in the "philosophical absolutization" of the natural sciences, and not in those sciences themselves, that one can detect "the ideological reflex of bourgeois society." [15] The understanding of natural phenomena according to laws dealing with the behavior of matter and expressed in mathematical terms is not itself a "reflection" or a "product" of modern society, but the assertion that nature must be represented exclusively in formal-mechanical terms and the claim that this system is the model of all scientific knowledge constitutes the sociologically determined aspect of the new positive science. These latter characteristics mark that science as originating in a form of society whose ethos is the domination of nature.

3. Critique

The conceptual transformation in natural philosophy which takes place gradually in the transition from the medieval to the modern Western world is very well known. Scheler is not concerned with describing the elements of that transformation once more, but rather with explaining it. How adequate is his account? Clearly the dramatic increase in the human ability to control, manage, and even alter the external environment results from the achievements of modern science and technology. Scheler tries to show how this was made possible by a particular procedure for the study of natural phenomena that eventually came to predominate over all others. In many respects his arguments that are summarized above are quite insightful. But the main question which should be raised is whether or not all this constitutes a proof for his thesis. Surely, it might be objected, we have increased not only our human power in the world but also our understanding of it, and perhaps the former is only a by-product of the latter. But this type of objection simply evades the problem and does not confront the basic issue involved: essentially, it shifts the argument to the level of purely linguistic considerations. In fact one can take the more "neutral" proposition—namely, that the modern natural sciences have increased our knowledge of the physical universe—for a starting point, but the question at issue here would remain the same: what are the *social consequences* of this increased understanding? Obviously from this perspective "growth of knowledge" and "mastery of nature" are not at all inconsistent.

Other authors after Scheler have attempted a further elaboration of his idea that modern science provides the basis for the human domination of nature. The chief defect in that work is the failure to carefully examine the concept of "domination" as it is applied in the notion of scientific knowledge as domination of nature. The failure originates with Scheler himself, for in his two long essays he nowhere subjects the crucial ambiguities in the idea of the domination of nature to the kind of rigorous examination which they deserve. Nevertheless his argument is indispensable, for it brings a host of problems to the surface and thus lays the groundwork for further clarification.

Let us assume for the sake of discussion that Scheler is correct in saying that the conceptual structure of modern science is "designed" for the mastery of nature. What does this tell us about the projected mastery of nature itself? In other words, if as Scheler claims this science seeks to dominate nature in the service of a human drive for power, what precisely is the nature of the power that is sought and how is it expressed? He is not very explicit on this point, and in order to answer these questions we have to focus on a few brief comments in his writings. But in fact the analysis of the relationship between science and domination depends upon the examination of this one basic point. In a passage referred to earlier Scheler maintains that the goal of the will to domination over nature sets itself the ideal task of framing a technical apparatus "through which man could lead and bind nature to whatever ends, be they useful or not, that he desired." Although Scheler mentions this point only in passing, it is actually the key to the whole issue. In order to understand the meaning of the attempted domination of nature, we have to decipher very carefully its relationship to those desired ends in the service of which it supposedly functions.

The will to mastery over nature that is externalized through science is, in Scheler's words, the "pragmatic motive" in the understanding of the world. This means that an inherently operational or instrumental framework (both theoretical and practical) guides the scientific-technological enterprise in the enlargement of the mastery of nature. What he apparently failed to notice, however, is that there is a decisive ambiguity involved in conceiving the scientific mastery of the world in pragmatic, operational, or instrumental terms. To be sure, Scheler correctly denies that immediate usefulness is the criterion of scientific work; rather, potential utility in the satisfaction of human desires is the hallmark of the pragmatic or operational approach, an increase in human power that would make possible an expansion of the realm of satisfaction at some point in the future. Thus when the scientific mastery of nature takes the form of instrumental or pragmatic reason, the putative goals which guide its work are the actual wants and desires of human beings. But among recent accomplishments are practices which threaten long-range ecological modifications and weapons whose use promises biological disaster. What is the relationship between these achievements, on the one hand, and human wants or desires on the other?

The ambiguity in the pragmatic conception arises out of the unrecognized fact that these wants and desires are, under the conditions of the struggle for existence in society, *con-*

tradictory. The error in the view of Scheler and others which presents the scientific mastery of the world as a pragmatic enterprise is that it leaves the realm of human goals and purposes unanalyzed. It is simply not sufficient to show that the scientific investigation of nature and its technological applications occur in an operational framework. The decisive question is, in what *specific* social context is it operational? If the context is—as is quite obviously the case—the worldwide struggle among social groups, then the violent social conflicts within and among nations subsist in a dialectical relationship with scientific-technological progress: each forces the other further along the path of its development. The qualitative leap in the attained mastery of the natural environment witnessed in the preceding few centuries is matched by a qualitative change in the human struggle for existence (from localized to global conflict) over that same period; each is simultaneously the cause and effect of the other. The objectives reflected in the will to domination over nature are not a simple collection of goals and purposes but rather an ensemble of partially contradictory ones. Scheler's basic error is to refer to a composite drive for power without further analyzing its conflicting components; the latter provide the key for the understanding of the historical dynamic of the will to domination.

Another expression of the same mistaken perspective is Scheler's portrayal of the difference between the medieval and the modern drive for power, the former expressing itself essentially in forms of domination over man, the latter in the domination of things. The suggested polarization between the two spheres is simply incorrect. Certainly the medieval forms of domination depended in large measure on the control of the means of subsistence (landed property) by a particular stratum in the population, and just as certainly the domination of man by man has not been rendered superfluous in modern society by the magic of a domination over things.

No doubt there is a difference between the forms of rule in the two periods, but Scheler's representation of it is difficult to accept.

The concept of *Herrschaftswissen* is incomplete, for until the meaning of "domination" in this context is made much clearer, we will not be able to comprehend the sense in which science and technology function as instruments for the winning of mastery over nature. Scheler tries to explain the conceptual transformations that herald the birth of modern science in terms of the heightened domination of nature which they made possible; he does not think he is describing only a philosophical movement, but also a change in the entire fabric of social life. It is not at all obvious, however, that the one had to follow necessarily from the other; in fact, exactly how the two are related remains problematical in the course of Scheler's exposition. He failed to see that mastery over nature as the expression of scientific and philosophical changes, and mastery over nature as a phenomenon related to the structure of social conflict in the modern period might not be identical.

The difference between these two aspects of mastery over nature may be stated briefly as follows. Modern science displays an increasing "mastery" in terms of its subject matter, namely, the conception of nature with which it works; in other words, its theoretical formulations achieve progressively greater generality and coherence—and thus greater rationality. At the same time these scientific accomplishments are translated into new possibilities for the satisfaction of human wants through the medium of technological progress, thereby giving rise to the popular belief that the human mastery over nature is being realized. And it is supposed that this accumulating mastery over nature lays the foundation for greater rationality in the structure and processes of society. The fundamental mistake in this conception is the expectation that the rationality of the scientific methodology itself (mastery of nature in the first sense) is "transferred" intact, as it were, to the social process and mitigates social conflict by satisfying human wants through the intensified exploitation of nature's resources (mastery of nature in the second sense). There is an ongoing dynamic relationship between these *two* forms of mastery over nature which must be depicted concretely; an attempt to do so has been made in the following pages.

Scheler described the purpose of modern science as the construction of a theoretical and practical scheme that will enable man to extract from nature resources sufficient to satisfy whatever desires he discovers in himself. The success of this project would represent mastery of nature. But if the pursuit of that scheme yields manifestly undesirable results, such as ecological and biological damage and the threat of nuclear annihilation, are these to be regarded as aspects of mastery too? Or should we admit in evidence only its beneficial consequences? Considered abstractly, modern scientific and technological developments would appear to represent in themselves a high degree of mastery over nature; however, if their desirable and undesirable features are as closely linked as they seem to be at present, the character of mastery over nature becomes exceedingly ambiguous. This ambiguity is not reflected in Scheler's concept of *Herrschaftswissen,* and therefore his account of the relationship between science and domination, although highly suggestive, is inadequate even on its own terms.

4. The Nature of Domination

Thus far in our discussion it has been possible to follow the customary practice by employing the terms mastery, domination, power, and conquest interchangeably, but at this point we can no longer do so. The vagueness of conventional usage now appears as an obstacle that must be overcome. The concepts of domination, power, and conquest must be rejected as wholly inappropriate when applied to the achievements of science with respect to nature, or at the very least they can be accepted in this context only with explicit qualifications. The notion of mastery, on the other hand, possesses different connotations which render it more suitable as a means of expressing what science contributes to the relationship between man and nature.

The stages in the historical development of the natural sciences have presented a progressively more coherent picture of the behavior of natural phenomena. In this sense, for example, Aristotelian, Newtonian, and relativistic physics are complementary, for in proceeding from various sets of assumptions each increased the human understanding of different classes or orders of natural phenomena. In themselves these sciences do not represent an increasing "power" over external nature; rather, their real significance with reference to the mastery of nature lies in the potential effect which they may have on human behavior within a peaceful social order embracing the human race as a whole. Under such conditions the rationality that is embodied in science may indeed serve as an instrument of man's self-mastery: to understand the world means in part to be at home in the world, to experience the harmony and order of its elements, to transcend that propensity to project alien and hostile forces onto natural events which arises out of insecurity and fear. Science shares this task of pacifying man's nature with religion, philosophy, and the arts, indeed with human culture in general.

We must realize, however, that this is a potential effect, not yet an actual one. Under existing conditions, where intense conflict rules the relations among individuals, social groups, and nations, this instrument is utterly powerless. The Aztecs believed that they were required to appease their gods by drenching the earth with the blood of their captives; today the same country that idolizes scientific progress and wields the most advanced technology also finds it necessary to sacrifice countless thousands of its foreign "enemies" to the insatiable demands of its "national security." [16] No matter how superior the scientific consciousness of a civilization may be, in itself this attribute cannot nourish rational human behavior so long as the violent struggle for existence persists.

The explanation for this dilemma is to be found in the fact that under present circumstances the specific results of research in the modern natural sciences cannot—for reasons which should be at least partially obvious—enter directly into the sphere of human practical life.[17] Information, techniques, and the reasoning of science "can be incorporated in the social life-world only through [their] technical utilization, as technological knowledge, serving the expansion of our power of technical control . . . [They] can only attain significance through the detour marked by the practical results of technical progress." [18] The concepts of power and domination, which do not make sense with respect to scientific knowledge itself, may be quite appropriately employed in connection with the technological applications of scientific knowledge. Advances in technology clearly enhance the power of ruling groups within societies and in the relations among nations; and as long as there are wide disparities in the distribution of power among individuals, social groups, and states, technology will function as an instrument of domination. The converse is likewise true: although new forms of science and technology may be associated with the struggle for hegemony waged by a rising social class (as is the case with the bourgeoisie in European civilization), in all forms of society characterized by the domination of a particular group neither science nor technology—strictly speaking—can serve as vehicles of general liberation, that is, the transition to a classless society.

The level of technological capability is an important factor in defining the form that social conflict will assume in any given period. This is why it is absurd to refer to "man's conquest of nature" or "man's domination of nature": the putative subject of this enterprise does not exist. "Man" as such is an abstraction which when employed in this manner only conceals the fact that in the actual violent struggles among *men* technological instruments have a part to play. The universality that is implied in the concept of man—the idea of the human race as a whole, united within the framework of a peaceful social order and finally determining its existence under the conditions of freedom—remains unrealized. On the other hand, so long as we restrict ourselves to the consideration of human potentialities, we can justly speak of the mastery of nature (that is, *human* nature) by *man* that may be achieved with the aid of science. As suggested earlier, the progress of science contributes to the promise of a unique kind of human self-mastery that may be possible under different social conditions, a kind of mastery that has none of the connotations implicit in the political analogies arising out of the notions of domination and conquest.

Scientific knowledge itself does not result in the "control" of external nature. The concept of power (at least in terms of its normal usage) is likewise inappropriate in this context, except insofar as the work of modern science is the indispensable prerequisite for all advanced technologies. In this respect, as Habermas argues, scientific knowledge presently affects human practice only through its technological applications, and these in turn constitute a crucial element in the social struggles among men.[19] In unison they may be said to be instruments of domination, but the real object of domination is not nature, but men. As Hegel showed in the famous section on "master and slave" in his *Phenomenology of the Spirit,* an essential feature of domination is the struggle for recognition of the master's authority. The necessary correlate of domination is the consciousness of subordination in those who must obey the will of another; thus properly speaking only other men can be the objects of domination.

If the idea of the *domination* of nature has any meaning at all, it is that by such means—that is, through the possession of superior technological capabilities—some men attempt to dominate and control other men. The notion of a common domination of the human race over external nature is nonsensical. This point can be understood best by examining what is signified by the word "nature" in relation to the mastery of nature through science.

6

SCIENCE AND NATURE

*Need and ingenuity have enabled man to
discover endlessly varied ways of mastering and making use of nature. . . . He
uses nature as a means to defeating
nature; the nimbleness of his reason
enables him to protect and preserve himself by pitting the objects of nature
against the natural forces which threaten
him and so nullifying them. Nature itself,
as it is in its universality, cannot be mastered in this manner however, nor bent
to the purposes of man.*

HEGEL, *The Philosophy of Nature*

1. Science and the Life-Crisis

In the middle of his intellectual career Max Scheler came
under the influence of Edmund Husserl's philosophy. Despite
the subsequent divergence of their interests and approaches,
there is an interesting affinity between the writings of Scheler's
that were discussed in the previous chapter and the outstanding work of Husserl's last period, *The Crisis of European
Sciences.* As we have seen, Scheler, in trying to relate a shift
in the scientific perspective on nature to broader social
changes occurring in the transition from medieval to modern
society, relies most heavily on classical European sociology
and philosophy of science. In a sense Husserl undertook a
much narrower study, and his exposition always remains

within the strict confines of his own "phenomenology." Nevertheless, the two contributions complement each other: whereas Scheler's concept of *Herrschaftswissen* requires us to scrutinize the meaning of domination in relation to the achievements of scientific knowledge, Husserl's book decisively raises the problem of understanding the meaning of nature in this same context.

To be sure *The Crisis of European Sciences* is not explicitly concerned with this problem, for its overt purpose is to serve as an introduction to phenomenological philosophy. Only twice in passing does Husserl mention the topic that we have been discussing. He writes:

> Along with his growing, more and more perfect cognitive power over the universe, man also gains an ever more perfect mastery over his practical surrounding world, one which expands in an unending progression. This also involves a mastery over mankind as belonging to the real surrounding world, i.e., mastery over himself and his fellow man, an ever greater power over his fate . . .[1]

Accordingly, only those aspects of his argument which are directly relevant to the present subject will be considered here. Although this procedure is less than fair to Husserl's work as a whole, the justification for it is that no other source so incisively exposes the question which must be framed at this point in our analysis of the idea of domination over nature.

Husserl himself published the first parts of his study as an article in 1936 (the remaining sections were edited from manuscripts after his death), at the time when—as noted earlier—the social function of science was a much-debated issue. This original essay was dedicated to an inquiry concerning the meaning of exact knowledge or science (*Wissenschaft*) in relation to human practical life, and in dramatic language the author announced in the heading of Part One that he intended to explore "the crisis of the sciences as [an] expression of the radical life-crisis of European humanity."

He referred to a change in the general public attitude toward the sciences and especially to the "feeling of hostility among the younger generation," and he suggested that this was caused by the opinion that the sciences had nothing to offer with respect to the most fateful questions which confront men, "questions of the meaning or meaninglessness of the whole of this human existence." He asked:

> Do not these questions, universal and necessary for all men, demand universal reflections and answers based on rational insight? In the final analysis they concern man as a free, self-determining being in his behavior toward the human and extrahuman surrounding world and free in regard to his capacities for rationally shaping himself and his surrounding world. What does science have to say about reason and unreason or about us men as subjects of this freedom? [2]

The heading of Section Two of the essay (in which this passage appears) identifies the crisis of science as the loss of its significance for life.

The function of Husserl's investigation is to discover how this loss of significance came about. In the latter half of the nineteenth century, he recalls, the situation had been far different: the most influential trains of thought had glorified the positive sciences and their fruits, and men generally felt that "prosperity" was intimately bound up with the continued success of the positive sciences in extending their domain and their methodology. In the twentieth century, however, Husserl finds that many writers speak of a condition of cultural crisis. The exact dimensions of this crisis are never clearly stated in his book; one is never sure whether Husserl is referring basically to a profound intellectual malaise of which the social problems of that time were partially the result, or whether he thinks that it had been social upheavals which had prompted the intellectual dilemma. Nor does Husserl deliver an opinion on the question of why this crisis became significant during the twentieth century.

In any case he agreed that it is of such a magnitude as to necessitate a reinvestigation of the foundations and the origins of modern scientific philosophy. What is at stake is the very concept of science itself, for as the above quotation shows, Husserl believed that we must seriously ask at this point whether the science which we have received and which we continue to practice has anything to say about the conditions under which the relationships of men to each other and to their natural environment could be determined in a free and rational way. When we ask about the concept of science, we immediately confront the decisive achievement of modern philosophy: the ideal of a universal science based on the model of mathematics and geometry. It is this ideal, whose chief innovators were Galileo and Descartes, which according to Husserl has determined the basic concept of science in modern Western history and which therefore must be investigated.

This model of science has two primary characteristics: (1) the separation of experience into subjective and objective factors—the ontological dimension; (2) the use of mathematics and geometry as the basic language of science—the methodological dimension. This science created a picture of the world in which a realm of eternally unchanging objects (the objects of mathematical physics) exists "behind" the fluctuating and deceptive realm of sense experience. The former realm is supposedly the one of true being, where a uniform matter exists as it really is, concealed behind the various manifestations of it that are perceived by the senses.

The basic point made by Husserl is that there are two "worlds" in the life of modern Western man, each radically different from the other; as he phrased it in an earlier manuscript, they are the discrete worlds of value objects and practical objects on the one hand, and the world of natural-scientific objects on the other. In the former we find things familiar to everyone—paintings, statues, gardens, houses, tables, clothes, tools, and so forth. In the latter is encountered another set of objects entirely, "ideal" objects which form the basis of mathematical operations.[3] Of course the two are related, most immediately in the fact that the process of experimental verification to which the mathematically formulated theories are subjected takes place in the familiar world. But Husserl contends that the nature of this relationship between the two worlds is highly problematical. This is so because there is no obvious connection between the truths of science and the orientation of life, both practical and intellectual, in everyday existence:

> The present European situation of a general breakdown of spiritual [*geistigen*] humanity changes nothing as far as the results of the natural sciences are concerned, and in their independent truth these results provide no impulse for a reformation of natural science. If such motives exist, they are concerned with the relation of these truths to scientific and extra-scientific humanity and its spiritual life.[4]

We live in the "subjective-relative" world whose features contrast sharply with those of the "objective-scientific" realm, and this dichotomy makes it difficult to understand the concrete unity of human action.

2. The Two Spheres of Human Activity

The familiar world is called by Husserl the "life-world" (*Lebenswelt*). It is the world of everyday experience; as used in this connection, "experience" does not have a primarily epistemological connotation, but rather refers to the regular encounter with the surrounding environment that occurs in the daily life of men and women. Experience in this world takes place on the level of ordinary intuition (a word similarly used in a nontechnical way), and it is always prescientific in the sense that it presupposes no special operations of any kind beyond the ordinary employment of human sensitivity and understanding. Every attempt at developing scientific or exact knowledge, of whatever kind, is a departure from the life-world, but only and always a partial departure, because in making such an attempt we remain of course actually existing in the everyday world. The theoretical coherence of any particular example of exact knowledge is in no way affected by this fact; but what is at stake is the human significance of every kind of exact knowledge, not its theoretical coherence. The meaning of any science must be investigated with reference to the fact that it arises out of human practice in the life-world, which encompasses the range of sciences at first as abstract possibilities. Thus when we inquire about the significance of a particular science in this context, we are not judging its alleged superiority or inferiority to any other type of science, but rather its meaning for activity in the life-world.

As far as the modern period is concerned, the other sphere that must be distinguished from the life-world is the activity of science, more precisely, the natural sciences. According to Husserl the mathematical-physical natural sciences have been for centuries the "self-evident" basis or model of exact science in Western civilization. The great value which has been placed upon its results and its methodology, and the outstanding influence which it has exerted upon the determination of the very concept of science itself, requires that the question concerning the human significance of science be addressed especially to this science. The investigation is directed not at any specific finding or findings attributed to it, but rather at the most fundamental operations which open the way to whatever results it might produce. In other words, what mode of abstraction guides its entire work, what principle of selectivity does it employ with respect to the natural environment that is the constant background of all our experience in the life-world?

The foundation stone for its specific mode of abstraction and selection is in Husserl's words the "mathematization of nature." Accordingly, the basic questions that must be framed are: "What is the meaning of this mathematization of nature? How do we reconstruct the train of thought which motivated it?" [5] This modern science attempts to define the universal structure of nature; and not only its structure, but also its process of causality (motion, interaction of physical bodies), is conceived in ideal or mathematical terms, as the famous laws of motion indicate. In short, there is presented a picture of abstract forms—matter conceived as identical with geometrical shape—interacting according to mathematically expressed formulas in geometrical space, as the universal condition of nature in all its manifestations. This idealization, for Husserl, "hides" its connection with the life-world; in other words, we cannot immediately understand how this universal conception of physical nature is related to ordinary experience.

The significance of the mathematization of nature, which is the central problem posed by Husserl, is twofold. In the first place, the relationship between experience in the life-world (common sense, intuited nature) and in the objective-scientific world (exact knowledge, mathematized nature) remains always unclear. The life-world is permanently "devalued" as the realm of purely subjective experience from the viewpoint of science, despite the fact that this is the world in which all human activity occurs. Secondly, the abstract-universal characteristics of the science based upon the mathematization of nature make it intrinsically impossible for that science to possess a direct relationship to specific goals formulated in human practice; the science can only make available through its technological applications a certain range of new possibilities for practical use. This means that such a science cannot transcend the purely technical level, that is to say, it cannot contribute to the formation of an objective basis for the judgments, choices, and valuations that must be made at every point in human practical life.

This is the resulting paradox: the methical abstractness of modern science, its discovery that all matter possesses a

uniform hidden structure and that the principles of its behavior are universally valid and can be expressed in mathematical formulas, is precisely the source of its astonishing
productivity in its ongoing interaction with technology. Necessarily, however, this very abstractness means that the scientific
understanding of nature and the scientific methodology—this
model of a silent, colorless universe of matter in motion—in
the final analysis remains mute in the theater of human
behavior. Husserl says that the actor in this drama (the scientist) never appears in the process of scientific investigation,
and "from this standpoint the rationality of the exact sciences
is of a piece with the rationality of the Egyptian pyramids." [6]
When this methodology makes its appearance in the social
sciences, for example, it advertises itself as "value-free" inquiry, that is, an approach wherein the researcher vigorously
attempts to divorce himself and his evaluations from the
material of his study. In illuminating the processes by which
choices and decisions are actually made under particular circumstances, these studies can result in the development of
techniques for controlling behavior; on the other hand, they
cannot aid in improving the quality of the choices that are
made. With respect to the control of both men and nature
we find ourselves in possession of ever more efficient means
for the accomplishment of ever more obscure ends.

In its original seventeenth-century formulation the new
scientific methodology was deliberately restricted in its application, concentrating exclusively on the investigation of external nature and relinquishing to religion the task of specify-
ing the objective rules of human conduct. This was true of
Bacon's endeavor, as we have seen, and equally so of Descartes's. The latter's *Discourse on Method* contained, together
with its revolutionary approach to scientific inquiry, a declara-
tion of obeisance to the established morality:

And finally, as it is not sufficient, before commencing to

rebuild the house which we inhabit, to pull it down and provide materials and an architect (or to act in this capacity ourselves, and make a careful drawing of its design), unless we have also provided ourselves with some other house where we can be comfortably lodged during the time of rebuilding, so in order that I should not remain irresolute in my actions while reason obliged me to be so in my [scientific] judgments, and that I might not omit to carry on my life as happily as I could, I formed for myself a code of morals for the time being . . .[7]

The first and most important maxim in this code was to obey the laws and customs of the country in which he found himself, and especially to adhere to the tenets of Christianity, for he felt it was "most expedient" to pattern his conduct according to that of the individuals with whom he would come into contact in daily life. But these temporary quarters for morality were never replaced with the solid edifice promised by Descartes, and André Gide could write of one of the characters in his *Lafcadio's Adventures* (*Les Caves du Vatican*): "The moral law which Descartes considered provisional, but to which he submitted in the meantime, until he had established the rules that should regulate his life and conduct hereafter, was the same law—its provisional powers indefinitely protracted—which governed Julius de Baraglioul."

Descartes's attack on the prevailing philosophical method of his day called into question the bases of *all* "objective" knowledge—implicitly, it shook the foundations of religion and ethics as well as those of the received science. He took refuge from the fateful consequences of his argument by fervently confessing his faith in the traditional religious dogmas, but this expedient gradually collapsed after his time. The limitations failed and (in Horkheimer's words) the new method was "absolutized": especially in positivist formulations, the methodology of natural science became the model for arriving at objective judgments of any kind, and whatever could not meet this test was confined to the realm of the merely subjective-relative. Already in this passage from Descartes there is an intimation of what Husserl would call the "life-crisis." It is in his acknowledgment that he could not remain "irresolute" in his actions—that is, that the ongoing concrete activity of practical life necessitated choices and valuations—while he was rebuilding the foundations of scientific knowledge.[8] What concealed the incipient crisis until the twentieth century was the continuing efficacy of religion, which managed to keep the structure of moral values in some state of repair despite the changing social and intellectual climate. But the overwhelming success of the marriage between industry and the new science, and the growing social authority of the novel scientific methodology, spelled inevitable defeat for the traditional scheme of religiously based ethics. For Bacon and his contemporaries religion had provided the means of understanding science as a human activity; to rephrase the point in Husserl's terminology, religion supplied the link uniting scientific activity with everyday action in the life-world. The failure of that link precipitated the crisis described earlier.

Husserl's contention that modern science "hides" its connection with the life-world would seem to contradict Scheler's viewpoint. For Scheler the kind of connection that Husserl regards as problematical is precisely what is most evident about this science, since one of its prime characteristics is the internal and necessary relationship between its theoretical structure and the accomplishment of practical goals. In concrete terms, the dynamic tie between the life-world and the scientific world is constituted by technology and technological progress: the development of more and more precise instrumentation is promoted by, and in turn itself promotes, continual advances both in opening up new problems and the possibility of their solution for the science and in providing new resources and techniques for the production of goods. But the resolution of the apparent contradiction is to be found in their disparate conceptions of means and ends. Scheler consigns the positive sciences and their theoretico-practical achievements to the sphere of practical objectives and sets beside them, as parallel and separate historical constants, the spheres of metaphysical and religious knowledge, where "final" ends and values are located. Thus for him the issue of the interconnection of these three strains does not arise. But in Husserl's view the connection between the life-world (where questions of values are crucial) and the scientific world is a critical problem because the apparent relationship between the two worlds as constituted by technology or technique says nothing about the inherent *rationality* of that relationship.

Scheler's scheme is basically unacceptable. Strictly speaking, there are no purely practical objectives, for every attempt to alter the means at our disposal also affects the perception and ranking of goals. For example, the attainment of better means for the satisfaction of material desires decisively undermines the social authority of the ascetic ideal and the longing for the hereafter, for release from the earthly vale of tears. Although it is sometimes tempting to say that science and technology procure for us a panorama of means to which our chosen ends may then be superadded in accordance with our will or whim, in fact such an idea is nonsensical, for means are not posited and attained in the abstract, isolated from any ends whatever. Husserl's search for a rational foundation for the interaction of the life-world and the scientific world is one way of expressing the necessity for a set of ends in which we can have at least as much confidence as we have in the means at our disposal.

3. The Bifurcation of Nature

Corresponding to the two spheres of human activity in modern life are two worlds of nature: intuited nature (*lebensweltliche Natur*) and scientific nature (*wissenschaftliche*

Natur), the experienced nature of everyday life and the abstract-universal, mathematized nature of the physical sciences. Ordinarily the former is not a "thematic" concern of consciousness, but rather subsists as the familiar background which to some extent binds together human experience on a universal scale despite all differences of culture and history. When something becomes a thematic concern, however, deliberate attention is focused on it; through a process of abstraction a certain aspect of experience becomes the subject of special interest, other aspects being simultaneously overshadowed and devalued. As previously indicated, this is the case with the modern scientific conception of nature: it selects a particular class of phenomena as its thematic field and excludes others.

On the one hand, we must concede that the reality that is apprehended in the modern natural sciences (the universal structure of matter and the laws of its behavior) was of course always "present"; on the other hand, only at a certain point in history did it actually become present *to* human consciousness and thus become comprehended as an aspect of nature. In this sense it is man's intellectual and practical activity which bestows upon nature the significance of a system of bodies in eternally lawful motion. Landgrebe phrases this point in phenomenological language as follows:

> The nature disclosed by natural science does not possess a being that is subsequently recognized by man; it rather receives this being by entering into the historical world of man and by being subjected to the experiments conducted by man. Only to the extent that nature can be subject to this kind of operative manipulation, can it be said to be "nature" in the sense of being the object of natural science, and only on this basis can natural science become an efficient tool for the technological domination of the world.[9]

In other words, nature *per se* is not the thematic object of the investigations pursued in the natural sciences, because there is simply no such thing. There are instead different perspectives on nature which are related to various types of human interests.[10] Others could be mentioned in addition to the two which Husserl isolates, for example the conception of nature developed in the Romantic movement and in nineteenth-century philosophy of nature. Although these cannot be entirely ignored, they have not (or not yet) entered the domain of concrete practical activity to a degree that requires us to rank them alongside the two we have been discussing as constitutive forces in experience.

We are now in a position to understand precisely why the ambiguity of the term "nature" is an essential factor in the problematical character of mastery of nature. Husserl's careful delineation of the two disparate realms of nature—the intuited, directly apprehended nature of universal experience in everyday life, and the idealized or mathematized nature of the modern scientific enterprise—raises a question which penetrates to the very core of the problem here under consideration: which nature is the object of mastery in the attempted mastery of nature?

Clearly the nature which is experienced in everyday life has been the object of mastery in every stage of human development. In general the control of nature in this sense has meant more or less complete disposition over the available natural resources of a particular region by an individual or social group and either partial or total exclusion of others from the benefits (and necessities of existence) available therein. In other words, under the conditions of the persistent social conflict that has characterized all forms of human society, the natural environment always appears either as already appropriated in the form of private property or else as subject to such appropriation. Accessibility to it is either denied or sharply restricted, actually or potentially.

Experience in the life-world includes, as part of the paramount reality of everyday life, conflict and struggle. The condition of the surrounding environment plays a vital role in that conflict: in the struggle between man and external nature, the latter is the source of both grief and satisfaction. From the point of view of man as a species, external nature appears as a reluctant and recalcitrant host; she does not willingly yield all her most precious fruits, and in order to possess them man must (to use Bacon's terminology) "vex," "hound," and "torment" her. But the benefits that are thus won at any particular time become the subject of more or less intense conflict, for in the struggles among men control of nature's resources is obviously a decisive weapon under most circumstances.

The dominant social groups in the various historical epochs indeed have achieved a notable degree of mastery over nature in the sense just described. Yet that mastery has been neither complete nor permanent, and in some respects the struggle among men for control of nature's resources has intensified during the modern period. As a result of the social movements of industrialization, urbanization, and capitalism in Western society (and to a lesser extent in the rest of the world) during the past few centuries, the majority of individuals have been deprived of all direct access to the material means of existence and have retained only their labor-power to be sold in the market under the threat of starvation or humiliating poverty; and yet even this extreme form of mastery over nature by the ruling strata did not succeed in repressing social conflict. On the contrary, such conflict steadily increased in scope and intensity, and gradually the social struggles that previously had been confined to relatively small regions became a unitary worldwide phenomenon which now occurs under the permanent threat of thermonuclear annihilation.

Such has been the fate of mastery over nature in modern times insofar as the surrounding environment and its resources are concerned. But the concept of mastery of nature is clearly meant to apply to the other world of nature as well. In fact modern science is supposed to represent domination over nature in a unique and highly developed form. Thus we must ask: what is the meaning of this mastery with reference to scientific nature?

To employ Bacon's phrasing once again, we may say that modern science expresses its mastery over nature in that it "takes off the mask and veil from natural objects, which are commonly concealed and obscured under the variety of shapes and external appearance" and deals with the "secrets" embedded in its hidden structure.[11] From the viewpoint of science the nature given in sense perception masks the underlying uniform structure of matter, and modern science's mastery consists in penetrating this disguise and identifying the characteristics of that structure. Considered from the opposite angle—from the viewpoint of life in the familiar world—the mastery of science is manifested in its ability to cast a "veil of ideas" (*Ideenkleid*) over the nature experienced in everyday existence, that is, to treat the phenomena of nature as if they were purely mathematical-geometrical objects.[12]

The scientific understanding of nature strives for the elaboration of a theoretical system in which all the axioms implicit in its conceptual foundation (the mathematization of nature) —or, to put the point in more familiar terms, all the laws of nature—have been fully unfolded, tested, and unified into a coherent picture. The idea of internal harmony, order, and regularity among the occurrences and behavior of natural phenomena, together with the notion of the universal applicability of the laws which govern them, act as heuristic principles of the intellectual disciplines that work with scientific nature: disharmony and internal inconsistency are signs of flaws, which should be eliminated, in the theoretical structure or in the experimental techniques, not in nature itself. These and other elements constitute the rationality of the scientific methodology arising out of the mathematization of nature, a rationality that has proved itself decisively in practice.

Mastery of nature in this sense means the increasing refinement of a theoretical scheme which explains that behavior consistently; and, while that explanation certainly need not be considered as a complete understanding of natural phenomena, it is an enduring contribution to whatever insight may be gained with respect to nature in general.

4. Scientific Rationality and the Social Milieu

There is an identifiable element of mastery, then, concerning both these worlds of nature, but it is also very different in each case. Indeed at first glance it would seem that as presented above they have nothing in common, and we might be able to remain content with this initial judgment were it not for the obvious historical connection between the progress of modern science and the multiplication of material benefits in everyday life that have been extracted from the natural environment—a connection constituted in practice by technology and industry. Thus we must recognize the fact that there is an ongoing interaction between these two worlds of nature and between the efforts to achieve mastery with respect to them. The task of describing the character of this interaction as carefully as possible must be undertaken so that some of the confusions apparent in contemporary literature—briefly summarized in Chapter One—may be dispelled. In the influential intellectual tradition originating with Bacon it has been assumed that mastery over nature considered as scientific-technological progress would be "automatically" transformed into mastery over nature considered as social progress (a reduction in the sources of social conflict). But if the two forms of mastery are quite different, as I have argued, then we must ask: what basis is there for assuming that mastery in the one domain can be translated into mastery in the other?

The developing element of mastery in the theoretical structure of modern natural science, its progress toward greater completeness and sophistication, is the fruit of its internal rationality. But that rationality necessarily remains bound to the domain of scientific nature and collapses in departing from it, because the conditions according to which that rationality first operates at all are established by the original idealization (the mathematization of nature) upon which it rests. The circumscription of the range of its application is the ransom exacted for its service. Any attempt to extend it further— to try to make it the basis for improving the conditions of social behavior, for example—is not itself an expression or an outcome of that rationality. Such attempts are facts pertaining to, or situations occurring in, the life-world, not the scientific world. Mastery of nature as the outcome of scientific rationality operating in the domain of scientific nature, when it is translated into the mode of mastery in an essentially different domain (practical action within the natural environment), cannot and does not preserve its character intact.

The best illustration for this point is provided by the persistent attempts to understand the nature and workings of society by means of a methodology borrowed from the natural sciences. The Enlightenment idea of a comprehensive "geometry of politics" or "social mathematics" has already been mentioned. In the nineteenth and twentieth centuries there have been repeated attempts on a lesser scale to represent aspects of the social process in the form of mathematically expressed principles or laws. Within definitely circumscribed limits these modest efforts have borne fruit, but the larger Enlightenment design has been consistently frustrated. This was inevitable, for the objective itself was misconceived; human behavior as a whole cannot adequately be comprehended if one strives to maintain the degree of abstractness necessary for mathematical symbolization at the same time. Hobbes was the only thinker who possessed sufficient courage and intellectual ruthlessness to construct a social theory strictly by analogy with the principles of the new natural-science methodology: for him society as well as external nature could be represented as a system of bodies in motion. It was a bold and startling analysis, but its ultimate limitations are clearly visible.[13] The internal rationality of the methodology cannot impose an order which is lacking in the subject matter itself (in this case, social behavior). Similarly, greater rationality in the scientific understanding of nature—even when it impinges so forcefully on practical life through dramatic technological applications —cannot insure by itself that greater rationality will prevail in the structure of the social process within which it has developed.

The two senses of mastery over nature described above have had an ongoing reciprocal impact on each other. The interaction is what enables us to regard it as a unified phenomenon, a dialectical unity of opposing elements whose inner tension is the source of its vitality. Considered as a whole, the significance of both poles, the pursuit of mastery through scientific rationality as well as its correlate in social life, can only be defined with reference to practical activity in the life-world. Husserl states: "If we cease being immersed in our scientific thinking, we become aware that we scientists are, after all, human beings and as such are among the components of the life-world which always exists for us, ever pregiven; and thus all of science is pulled, along with us, into the—merely 'subjective-relative'—life-world." [14] *Mastery of nature considered as scientific rationality submits to the conditions that define mastery of nature in the prevailing historical and social world.* To be sure, the former does not merely "submit." Its achievements, although constituting mastery of nature in a different sense, are immensely influential in setting the limits of the options available for the pursuit of mastery in the life-world. In turn the latter, as a result of both the attained and the promised accomplishments of scientific rationality, influences the progress of the former by gradually establishing its priority as a thematic concern over other competing interests, such as religion. Similarly, the allocation of social resources, which is determined in the struggle of competing interests in social life, affects the rate of development in science.

We may now be able to realize why the idea of mastery of nature is the subject of so many confusing remarks and contradictory interpretations. It is at one and the same time a revealing and a concealing notion, and its deceptive power arises from the fact that its meaning appears so self-evident. In reality the historical phenomena that it denotes have no stable core; rather their point of intersection represents an unstable, ever-shifting resolution of forces. Thus while it identifies certain important interrelationships among historical tendencies, it also masks the element of inherent flux in them. Most interpreters of the idea concentrate on either one or the other of the two poles described above, and therefore their analyses are invariably misleading. Only when the two are discussed in relation to each other is it possible to limn the full dimensions of the problem. An illustration representing an extreme case might help to clarify this point, although it should be borne in mind that this is not a typical instance and is chosen because in such cases the contradictory features of the idea are thrown into sharp relief.

Even under relatively primitive circumstances men wage a kind of warfare against the natural environment by way of response to the pressures of the struggle for existence. An example is the deliberate burning of great sections of forested land on the African continent for the purpose of increasing the area available for agriculture, a deed which often recoils disastrously on the unwitting perpetrators because of the resulting natural imbalance in the environment. But this kind of situation is not confined to so-called "primitive" frameworks. The persistence of the struggle for existence reproduces such contradictions even at the highest attained levels of rationality. At these levels, where the duality of intuited nature and scientific nature is present, the technological capabilities based upon the latter vastly expand human capacities to conduct warfare against the natural environment under the conditions of intensified social conflicts. The confrontation of guerrilla operations and counterinsurgency techniques is the most dramatic contemporary instance. The natural environment of the everyday world serves as one of the guerrilla's chief supports: the mantle of darkness and the cover of the jungle canopy help to offset the conventional military superiority of his opponent. And in the extreme case, instruments made possible by the most advanced scientific and technological accomplishments are employed against the natural environment itself in efforts to defeat the guerrilla. Chemical warfare disguised under the euphemism of "defoliation" simply removed vast portions of jungle growth in Vietnam, and the military strategists proposed the orbiting in space of gigantic solar reflectors which would have turned night into day for that same country.[15]

The effort to understand mastery of nature forces us to explore its implications in the context of the bitter social conflicts that are gradually coalescing into an overall global drama. In those conflicts the accelerating pace of technological change plays a significant role. Unfortunately, any examination of contemporary sociological literature will show that the social consequences of technological progress are as yet only poorly adumbrated. Therefore in the next stage of our discussion here we must outline the function of technology in the mastery of nature.

7

TECHNOLOGY AND DOMINATION

> *Technology reveals the active relation of man to nature, the immediate process of production of his life, and thereby also his social life-relationships and the cultural representations that arise out of them.*
>
> MARX, *Capital*

1. Introduction

In the preceding pages technology has been described as the concrete link between the mastery of nature through scientific knowledge and the enlarged disposition over the resources of the natural environment which supposedly constitutes mastery of nature in the everyday world. Normally the rubric "conquest of nature" is applied to modern science and technology together, simply on account of their manifest interdependence in the research laboratory and industry. When they are considered in isolation, as two related aspects of human activity among many others, the fact of their necessary connection must indeed be recognized if their progress in modern times is to be understood. But it does not follow automatically that they function as a unity with respect to the mastery of nature, since they are not identical with the latter: mastery of nature develops also in response to other aspects in the social dynamic, for example the process whereby new human needs are formed, and therefore its meaning with respect to technology may be quite different than it is in the case of science.

I have tried to show that nature *per se* is not the object of mastery, that instead various senses of mastery are appropriate to various perspectives on nature. If this proposition is correct, then the converse is likewise true, namely, that mastery of nature is not a project of science *per se* but rather a broader social task.[1] In this larger context technology plays a far different role than does science, for it has a much more direct relationship to the realm of human wants and thus to the social conflicts which arise out of them. This is what Marx meant in referring to the "immediate" process of production in which technology figures so prominently—the direct connection between men's technical capacities and their ability to satisfy their desires, which is a constant feature of human history and is not bound to any specific form of scientific knowledge. On the other hand, science, like similar advanced cultural formations (religion, art, philosophy, and so forth), is indirectly related to the struggle for existence: in technical language, these are all *mediated* by reflective thought to a far greater extent. Of course this by no means implies that they lack a social content altogether, but only that it is present in highly abstract form and that by virtue of their rational impulse they transcend to some extent the specific historical circumstances which gave them birth. Therefore scientific rationality and technological rationality are not the same and cannot be regarded as the complementary bases of something called the domination of nature.

The function of scientific rationality in the mastery of nature has already been indicated, and the character of technological rationality in the same context must now be explored. Two considerations are especially relevant to the discussion. In the first place, the immediate connection of technology with practical life-activity determines *a priori* the kind of mastery over nature that is achieved through technological development: caught in the web of social conflict, technology constitutes one of the means by which mastery of nature is linked to mastery over man. Secondly, the employment of technological rationality in the extreme forms of social conflict in the twentieth century—in weapons of mass destruction, techniques for the control of human behavior, and so forth—precipitates a crisis of rationality itself; the existence of this crisis necessitates a critique of reason that attempts to discover (and thus to aid in overcoming) the tendencies uniting reason with irrationalism and terror. These two themes have been brilliantly presented in the work of the contemporary philosopher Max Horkheimer.

The attempt to understand the significance of the domination of nature is a problem with which Horkheimer has been concerned during his entire career: one can find scattered references to it in books and essays of his spanning a period of forty years. In my estimation his analysis, although quite unsystematic, contains greater insight into the full range of the problem under discussion here than does any other single contribution, although one can of course find many affinities between his work and that of others. For example, his description of the idea of nature which guides the investigations of the modern natural sciences is almost identical with Scheler's. In addition, the fundamental question which he poses is similar to the one raised by Husserl (although from a very different philosophical point of view) in *The Crisis of European Sciences:* what is the concept of rationality that underlies modern social progress? He also shares with Husserl the conviction that the social dilemmas related to scientific and technological progress have reached a critical point in the twentieth century. Horkheimer describes this situation as one in which "the antagonism of reason and nature is in an acute and catastrophic phase." [2]

2. The Critique of Reason

Horkheimer follows Nietzsche's pathbreaking thought and argues that the domination of nature or the expansion of human power in the world is a universal characteristic of human reason rather than a distinctive mark of the modern period:

> If one were to speak of a disease affecting reason, this disease should be understood not as having stricken reason at some historical moment, but as being inseparable from the nature of reason in civilization as we have known it so far. The disease of reason is that reason was born from man's urge to dominate nature. . . . One might say that the collective madness that ranges today, from the concentration camps to the seemingly most harmless mass-culture reactions, was already present in germ in primitive objectivization, in the first man's calculating contemplation of the world as a prey.[3]

Again like Nietzsche he finds the first clear expression of this will to power in the rationalist conceptions of ancient Greek philosophy, conceptions which determined the predominant course of subsequent Western philosophy. The concept (*Begriff*) itself, and especially the concept of the "thing," serves as a tool through which the chaotic, disorganized data of sensations and perceptions can be organized into coherent structures and thus into forms of experience that can give rise to exact knowledge. The deductive form of logic which emerges in Greek philosophy reinforces this original tendency and raises it to a much higher level. The deductive form of thought "mirrors hierarchy and compulsion" and first clearly reveals the social character of the structure of knowledge: "The generality of thought, as discursive logic develops it— domination in the realm of concepts—arises on the basis of domination in reality." The categories of logic suggest the power of the universal over the particular and in this respect they testify to the "thoroughgoing unity of society and domination," that is, to the ubiquity of the subjection of the individual to the whole in human society.[4]

Horkheimer differs from Nietzsche in attempting to distinguish two basic types of reason. Although in themselves all structures of logic and knowledge reflect a common origin in the will to domination, there is one type of reason in which this condition is transcended and another in which it is not: the former he calls objective reason, the latter, subjective reason.[5] The first conceives of human reason as a part of the rationality of the world and regards the highest expression of that reason (truth) as an ontological category, that is, it views truth as the grasping of the essence of things. Objective reason is represented in the philosophies of Plato and Aristotle, the Scholastics, and German idealism. It includes the specific rationality of man (subjective reason) by which man defines himself and his goals, but not exclusively, for it is oriented toward the whole of the realm of beings; it strives to be, as Horkheimer remarks, the voice of all that is mute in nature. On the other hand, subjective reason exclusively seeks mastery over things and does not attempt to consider what extra-human things may be in and for themselves. It does not ask whether ends are intrinsically rational, but only how means may be fashioned to achieve whatever ends may be selected; in effect it defines the rational as that which is serviceable for human interests. Subjective reason attains its most fully developed form in positivism.

The two concepts do not exist for Horkheimer as static historical constants. Objective reason both undergoes a process of self-dissolution and also succumbs to the attack of subjective reason, and this fate in a sense represents necessary historical progress. The conceptual framework and hierarchy established by objective reason is too static, condemning men (as the subjects of historical change) to virtual imprisonment in an order which valiantly tries to maintain its traditional foundations intact. By the seventeenth century, for example, the combination of Aristotelianism and Christian dogma in late medieval philosophy had become intellectually sterile, a system in which the repetition of established formulas was substituted for original thought. The declining period of great philosophical systems is characterized by the increasingly effective onslaught of skepticism, such as that which struck Greek metaphysics in Hellenistic times and scholastic philosophy in the sixteenth century. The skeptics represent movements of "enlightenment"—a recurring pattern of which the eighteenth-century French Enlightenment is the most famous example— wherein thinkers undermine concepts and dogmas that were once vital but have subsequently ossified, often in the process turning into ideological masks for the material interests of certain social groups.

But in its later stages the movement of enlightenment reveals its own internal contradictions, represented by Horkheimer and Adorno in their famous notion of the "dialectic of enlightenment." [6] What marks the general program of enlightenment as a unitary phenomenon despite its various historical guises is the "demythologization of the world." In its earliest period it combatted the traditional religious mythologies (for example, in Greek civilization); in the modern West it takes the form of a struggle against mystification in religion and philosophy, and in its most advanced stages—as in positivism—it carries this campaign to the heart of conceptual thought itself, finally upholding the position that only propositions conforming to one particular notion of "verifiable knowledge" have any meaning at all. The sustained effort of demythologizing in modern times ends by stripping the world of all inherent purpose. Nature, for example, appears to scientific thought only as a collection of bodies in eternally lawful motion, and the social reflection of this scientific vision is the idea of a set of natural laws of economic behavior which blindly follow their established course and which exhibit no inherent rationality. The consequence of this view is to set the relationship of man and the world inescapably in the context of domination: man must either meekly submit to these natural laws (physical and economic) or attempt to master them; for since they possess no purpose, or at least none that he can understand, there is no possibility of reconciling his objectives with those of the natural order.

The dialectical nature of enlightenment becomes clear only at this advanced level. All other purposes having been driven out of the world, only one value remains as evident and fundamental: self-preservation. This is sought through mastery of the world to assure the self-preservation of the species and, within the species, through mastery of the economic process to assure the self-preservation of the individual. Yet the puzzling fact remains that adequate security (as the goal of self-preservation) is never attained, either for the species or the individual, and sometimes seems to be actually diminishing for both. Thus the struggle for mastery tends to perpetuate itself endlessly and to become an end in itself.

In the course of enlightenment the predominant function of reason is to serve as an instrument in the struggle for mastery. Reason becomes above all the tool by which man seeks to find in nature adequate resources for self-preservation. It separates itself from the nature given in sense perception and finds a secure point in the thinking self (the *ego cogito*), on the basis of which it tries to discover the means for subjecting nature to its requirements. In the new natural philosophy of the seventeenth century, this procedure is adopted as the theme of science, and as the principal mode of behavior through which the mastery of nature is pursued, science assumes an increasingly influential role in society. For Horkheimer the attributes of the modern scientific conception of nature which predispose it for the purposes of mastery are, in part: the principle of the uniformity of nature, the inherent technological applicability of its findings, the reduction of nature to pure "stuff" or abstract matter through the elimination of qualities as essential features of natural phenomena, and especially the primacy of mathematics in the representation of natural processes.[7]

Horkheimer does not present this picture of science in isolation, but rather tries to understand what complementary conditions are necessary in order for that science to become, as it has, a historical reality of great dimensions. He contends that the "mastery of inner nature" is a logical correlate of the mastery of external nature; in other words, the domination of the world that is to be carried out by subjective reason presupposes a condition under which man's reason is already master in its own house, that is, in the domain of human nature. The prototype of this connection can be found in Cartesian philosophy, where the ego appears as dominating internal nature (the passions) in order to prevent the emotions from interfering with the judgments that form the basis of scientific knowledge. The culmination of the development of the transcendental subjectivity inaugurated by Descartes is to be found in Fichte, in whose early works "the relationship between the ego and nature is one of tyranny," and for whom the "entire universe becomes a tool of the ego, although the ego has no substance or meaning except in its own boundless activity."[8]

In the social context of competition and cooperation the abstract possibilities for an increase in the domination of nature are transformed into actual technological progress. But in the ongoing struggle for existence the desired goal (security) continues to elude the individual's grasp, and the technical mastery of nature expands as if by virtue of its own independent necessity, with the result that what was once clearly seen as a means gradually becomes an end in itself:

> As the end result of the process, we have on the one hand the self, the abstract ego emptied of all substance except its attempt to transform everything in heaven and on earth into means for its preservation, and on the other hand an empty

nature degraded to mere material, mere stuff to be dominated, without any other purpose than that of this very domination.[9]

On the empirical level the mastery of inner nature appears as the modern form of individual self-denial and instinctual renunciation required by the social process of production. For the minority this is the voluntary, calculating self-denial of the entrepreneur; for the majority, it is the involuntary renunciation enforced by the struggle for the necessities of life.

The crucial question is: what is the historical dynamic that spurs on the mastery of internal and external nature in the modern period? Two factors shape the answer. One is that the domination of nature is conceived in terms of an intensive exploitation of nature's resources, and the other is that a level of control over the natural environment which would be sufficient (given a peaceful social order) to assure the material well-being of men has already been attained. But external nature continues to be viewed primarily as an object of potentially increased mastery, despite the fact that the level of mastery has risen dramatically. The instinctual renunciation—the persistent mastery and denial of internal nature—which is required to support the project for the mastery of external nature (through the continuation of the traditional work-process for the sake of the seemingly endless productive applications of technological innovations) appears as more and more irrational in view of the already attained possibilities for the satisfaction of needs.

Horkheimer answers the question posed in the preceding paragraph as follows: "The warfare among men in war and in peace is the key to the insatiability of the species and to its ensuing practical attitudes, as well as to the categories and methods of scientific intelligence in which nature appears increasingly under the aspect of its most effective exploitation." [10] The persistent struggle for existence, which manifests itself as social conflict both within particular societies and also among societies on a global scale, is the motor which drives the mastery of nature (internal and external) to ever greater heights and which precludes the setting of any *a priori* limit on this objective in its present form. Under these pressures the power of the whole society over the individual steadily mounts and is exercised through techniques uncovered in the course of the increasing mastery of nature. Externally, this means the ability to control, alter, and destroy larger and larger segments of the natural environment. Internally, terroristic and nonterroristic measures for manipulating consciousness and for internalizing heteronomous needs (where the individual exercises little or no independent reflective judgment) extend the sway of society over the inner life of the person. In both respects the possibilities and the actuality of domination over men have been magnified enormously.

As a result of its internal contradictions, the objectives which are embodied in the attempted domination of nature are thwarted by the enterprise itself. For the mastery of nature has been and remains a social task, not the appurtenance of an abstract scientific methodology or the happy (or unhappy) coincidence of scientific discovery and technological application. As such its dynamic is located in the specific societal process in which those objectives have been pursued, and the overriding feature of that context is bitter social conflict. Even in those nations where the fruits of technical progress are most evident,

> despite all improvements and despite fantastic riches there rules at the same time the brutal struggle for existence, oppression, and fear. That is the hidden basis of the decay of civilization, namely that men cannot utilize their power over nature for the rational organization of the earth but rather must yield themselves to blind individual and national egoism under the compulsion of circumstances and of inescapable manipulation.[11]

The more actively is the pursuit of the domination of nature undertaken, the more passive is the individual rendered; the greater the attained power over nature, the weaker the individual vis-à-vis the overwhelming presence of society.

3. The Intensification of Social Conflict

Horkheimer's theory ties together three features of human history: domination of nature, domination over man, and social conflict. The contemporary literature cited in Chapter One recognizes only a puzzling affinity between the first two. But if Horkheimer is correct, there is a *necessary* connection between them under present circumstances: the third factor is the binding element. This theory diverges substantially from the accepted wisdom, and therefore a clear statement of it is essential. Horkheimer's argument can be rephrased in the form of a theoretical model that displays the dynamic interaction of all three factors. In this paradigm they appear only at the highest level of abstraction, for its purpose is to suggest a conceptual framework that will uncover the hidden nexus among actual historical events. Strictly speaking, it is an hypothesis whose adequacy must be tested and judged in relation to its ability to explain a wide range of empirical data.

So long as the material basis of human life remains fixed at a relatively low level and bound to premechanized agricultural production, the intensity of the struggle for existence fluctuates between fairly determinate limits. The material interdependence of men and women in different areas under such conditions is minimal, and the lack of any appreciable control over the natural environment also constricts efforts to extend the hegemony of particular groups permanently beyond their local borders. Political domination within and among societies is everywhere at work, to be sure, but it is also severely limited in scope. Slowness of communications and transportation hampers the exercise of centralized authority, which outside the area of its immediate presence is restricted to intermittent displays of its might; the daily struggle for the requirements of life normally occurs on a local basis. As mentioned earlier, in all forms of society characterized by class divisions the natural environment surrounding the individual in everyday life appears as actually or potentially in another's domain. The fear of being denied access to the means of survival is a determining aspect of the relationship between man and external nature in the evolution of society. But in the premechanized agricultural economy both ruler and ruled are subject to the parsimonious regime of nature: the comparatively low productivity of labor, the paucity of the economic surplus, and the small accumulated reserves of commodities generally check the designs of empire or at least render both domestic and imperial authority highly unstable.

The link between the struggle for existence and control of the natural environment is illustrated best by the fact that the intensity of the possible exploitation of human labor is directly dependent upon the attained degree of mastery over external nature. Here the decisive step has been the coming of industrial society: the machine and the factory system have expanded enormously the productivity of labor and consequently the possible margin of its exploitation. Thus the heightened mastery of external nature reveals its social utility in the mounting productivity of labor resulting from the technological applications of scientific knowledge in the industrial system. But why does there also occur a qualitative leap in the intensity of social conflict? In the first place, the economic surplus, which in class-divided societies is appropriated as private property, becomes so much larger and opens new opportunities for the development and satisfaction of needs, both material and cultural; consequently disposition over this surplus becomes the focus of greater contention. Second, certain types of natural resources (for example, coal and oil), available only in specific areas, become essential ingredients for the productive process. An adequate supply of these resources must be assured, and so the commercial tentacles of the productive unit must expand, until in some instances it draws upon supplies extracted from every corner of the planet. Inasmuch as every productive unit becomes dependent upon its sources of raw materials, every actual or potential denial of access to them represents a threat to the maintenance of that unit and to the well-being of its beneficiaries. Since obviously no equitable distribution of the world's natural resources has been agreed upon, the effect of that widened dependency is to magnify the scope of conflict.

The imbalance among existing societies in the attained level of mastery over the external environment acts as a further abrasive influence. The staggering growth in the destructiveness of weapons and in the capabilities of the "delivery systems" for them aggravates the fears and tensions in the day-to-day encounters among nations, whether or not those weapons are ever actually employed.[12] The most favored nations in this regard may wreak havoc anywhere on the globe, and those less fortunate must either hope for parity or expect to suffer repeated ignominy. The fact that every social order must fear the depredations not only of its immediate neighbors but potentially of every remote country—a condition arising out of increased mastery of nature accomplished in the context of persistent social conflict—alters the stakes in the dangerous game of human rivalry.

A fourth contributory factor may also be mentioned, namely, the extension of the struggle to the realm of the spirit through intensive propaganda (both domestic and foreign) and the manipulation of consciousness. The dystopian novelists are fond of exhibiting this feature of their prognosis, but we have had already an ample foretaste of it in experience. Finally, the rising material expectations of populations grown accustomed to an endless proliferation of technological marvels have a decisive impact. In this respect, mastery of nature without apparent limit becomes the servant of insatiable demands made upon the resources of the natural environment, that is, demands for the transformation of those resources into a vast realm of commodities. Perhaps they can be met—even on a universal scale, for all men. Yet if every level of gratification for material wants merely serves to elicit a more elaborate set of desires, the competitiveness and isolation among individuals that underlies the psychology of consumer behavior will continue to feed the sources of conflict.

Through the attempted conquest of nature, therefore, the focus of the ongoing struggle of men with the natural environment and with each other for the satisfaction of their needs tends to shift from local areas to a global setting. For the first time in history the human race as a whole begins to experience particular clashes as instances of a general worldwide confrontation; apparently minor events in places far removed from the centers of power are interpreted in the light of their probable effect on the planetary balance of interests. The earth appears as the stage-setting for a titanic self-encounter of the human species which throws into the fray its impressive command over the forces of nature, seemingly determined to confirm the truth of Hegel's dictum that history is a slaughterbench. The idea of man as a universal being, one of the great achievements of philosophical and religious thought, is refracted through the prism of universal conflict and realized in a thoroughly distorted form.

The cunning of unreason takes its revenge: in the process of globalized competition men become the servants of the very instruments fashioned for their own mastery over nature, for the tempo of technological innovation can no longer be controlled even by the most advanced societies, but rather responds to the shifting interplay of worldwide forces. Entire peoples and their fragile social institutions, designed for far different days, are precipitously sucked into the maelstrom. After centuries of benign neglect the population of the Sudan, for example, is suddenly offered advice and martial gifts by a medley of agents from Egypt, the Soviet Union, the United States, Libya, the German Democratic Republic, and Israel, while its neighbors in Ethiopia battle each other with American and Chinese armaments.

The intensification of conflict through the technological mastery of nature thus precipitates a search for new techniques for the exercise of political domination among men. The heightening struggle sets men against each other more desperately and requires that measures be taken to contain the escalating pressures of conflict. The terroristic devices of totalitarian rule are well-known instances, but these measures are present in nonterroristic forms as well. Under monopoly capitalism, for example, the technological mastery of nature turns into mastery over men through the manipulation of needs; but the clearest cases are to be found in the context of the real or presumed threats to imperial rule that have arisen in the so-called "third world." As befits what the sociologists have labeled a "knowledgeable society," social scientists in the United States have been drafted into the counterinsurgency effort: the CIA agent travels in company with his academic *éminence grise.* The culture and social organization of the most remote tribes have been intensively studied so that techniques of control appropriate to those peoples should be available if required.[13]

Political mastery over human nature in all its diverse cultural forms is sought in response to the intensified social conflict which in turn depends in part on a growing mastery over external nature. Certainly the last-mentioned factor is not the only element that figures prominently in this conflict, but it is certainly one of the most significant. Also, the relationships among these three are never unidirectional; reciprocal interaction always exists. The overriding reality of social antagonism in which they are united destroyed the dream of the seventeenth and eighteenth centuries that scientific and technological rationality could mold a harmonious society. In Habermas's words: "The substance of domination is not dissolved by the power of technical control." [14] It is equally true that the advanced techniques of political domination can only repress temporarily, not eliminate, the manifestations of social conflict; such techniques are no more efficacious than were the more primitive variants in securing permanently the hegemony of particular groups or nations. A condition of permanent warfare among regional superstates on the model of Orwell's *1984* would seem to be about the best that the captains of empire could hope for, but it is unlikely that even this degree of domination can be achieved.

Are there countervailing tendencies at work against this process of magnified conflict and domination? Socialism promised to bring the social relations of men—and therewith also the relation of man to nature—under rational control. Ironically, the very success of socialism in practice has resulted in that uneasy confrontation of powerful social systems which has placed the whole earth under the sword of Damocles. But whether it is known by this or any other name, the effort to frame institutions capable of subjecting the global social dynamic to the collective control of free individuals now represents an insurmountable necessity for the human race as a whole. Having allowed free rein to their talents in mastering external nature, men must now learn how to domesticate their own ingenuity.

The attempt to realize the traditional promises of socialism takes place within the established socialist societies as well as in the ongoing struggle for socialism elsewhere. Naturally, collectivization in production itself provides no guarantee that the social decision-making process will function rationally; the crucial problem everywhere is to eliminate relations of domination and subordination as rapidly as possible. The essential task is clear, but the situation remains in the balance today, and all prediction—including the learned divinations of the futurologists—is unreliable. The impact of certain prospective events, especially the imminent recognition of China's great civilization and enormous population as a decisive factor in world politics, cannot be gauged at present.

Once this framework of collective rational control is established, technology will be liberated from its all-too-effective service in the cause of human conflict. Until that time, however, we remain victims of a dilemma whereby every outstanding victory in the scientific and technological mastery of nature entails the real possibility of an equally great catastrophe.

4. The Revolt of Nature

The growing domination of men through the development of new techniques for mastering the natural environment and for controlling human nature does not go unresisted. Horkheimer analyzes the reaction to it under the heading of the "revolt of nature," a brilliant and original conception that has never received the attention it deserves.[15] The revolt of nature means the rebellion of human nature which takes place in the form of violent outbreaks of persistently repressed instinctual demands. As such it is of course not at all unique to modern history, but is rather a recurrent feature of human civilization. What is new in the twentieth century, on the other hand, is the fact that the potential scope of destructiveness which it entails is so much greater.

There are many reasons for this. The most fundamental is the fact that—at least in the industrially advanced countries—the traditional grounds for the repression of instinctual demands have been vitiated but yet continue to operate: the denial of gratification, the requirements of the work process, and the struggle for existence persist almost unchanged despite the feasibility of gradually mitigating those conditions by means of a rational organization of the available productive forces. The reality principle which has prevailed throughout history has lost most of its rational basis but not its force, and this introduces an element of irrationality into the very core of human activity: "Since the subjugation of nature, in and outside of man, goes on without a meaningful motive, nature is not really transcended or reconciled but merely repressed." [16] The denial of instinctual gratification—the subjugation of internal nature—is enforced in the interests of civilization; release from this harsh regime of the reality principle was to be found in the subjugation of external nature, which would permit the fuller satisfaction of instinctual demands while preserving the order of civilization. The persistence of social conflict thwarts that objective, however, and prompts a search for new means of repressing the sources of conflict in human nature. Security is sought in the power over external nature and over other men, a power that seems possible on the basis of the remarkable accomplishments of scientific and technological rationality, but the need for security, arising always afresh out of the irrational structure of social relations, is never appeased. The dialectic of rationality and irrationality feeds the periodic outbursts of destructive passions with ever more potent fuels.

Secondly, the revolt of human nature, directed against the structure of domination and its rationality, is proportional in intensity to that of the prevailing domination itself. Greater pressures produce correspondingly more violent explosions; the magnified level of domination in modern society, achieved in respect to both external and internal nature (as we have seen, both make their effects felt in everyday social life despite their differing immediate objects), is also a measure of the heightened potential of the revolt of nature. Thirdly, in recent times this revolt itself has been manipulated and encouraged by ruling social forces as an element in the struggle for sociopolitical mastery. Horkheimer refers here to fascism, which cultivates the latent irrationalities in modern society as material to be managed by rational techniques (propaganda, rallies, and so forth) in the service of political objectives.

The idea of the revolt of nature suggests that there may be an internal limit within the process of enlarging domination that was outlined above. Certainly, at every level of technological development the irrationalities present in the structure of social relations have prevented the realization of the full benefits that might have been derived from the instruments (including human labor) available for the exploitation of nature's resources. The misuse, waste, and destruction of these resources at every stage is at least partially responsible for the continued search for new technological capabilities, as if the possession of more refined techniques could somehow compensate for the misapplication of the existing ones. And because of the lasting institutional frameworks through which particular groups control the behavior of others, the new techniques are utilized sooner or later in the service of domination.

Yet it does not seem possible that this process can continue indefinitely, for at the higher levels the gap between the rational organization of labor and instrumentalities on the one hand, and the irrational uses to which that organization is put on the other, widens to the point where the objectives themselves are called into question. The problem is not only that the degree of waste and misuse of resources has increased enormously, but also that the implements of destruction now threaten the biological future of the species as a whole. This is the point beyond which the nexus of rational techniques and irrational applications ceases to have any justification at all; it represents the internal limit in the exercise of domination over internal and external nature, to exceed which entails that the intentions are inevitably frustrated by the chosen means.

The purpose of mastery over nature is the security of life—and its enhancement—alike for individuals and the species. But the means presently available for pursuing these objectives encompass such potential destructiveness that their full employment in the struggle for existence would leave in ruins all the advantages so far gained at the price of so much suffering. In the intensified social conflicts of the contemporary period, and especially in the phenomenon of fascism, Horkheimer sees this dialectic at work, and this is what he has tried to describe in the notion of the revolt of nature. The use of the most advanced rational techniques of domination over external and internal nature to prevent the emergence of the free social institutions envisaged in the utopian tradition represents for him the blind, irrational outbreak of human nature against a process of domination that has become self-destructive.

In the interval since Horkheimer first presented this notion a related aspect of the problem has been recognized: in a different sense the concept of the revolt of nature may be applied in relation to ecological damage in the natural environment. There is also an inherent limit in the irrational exploitation of external nature itself, for under present conditions the natural functioning of various biological ecosystems is threatened. It is possible that permanent and irreversible damage to some parts of the major planetary ecosystems may have already occurred; the consequences of this are not yet clear.[17] If it is the case that the natural environment cannot tolerate the present level of irrational technological applications without suffering breakdowns in the mechanisms that govern its cycles of self-renewal, then we would be justified in speaking of a revolt of external nature which accompanies the rebellion of human nature.

The dialectic of reason and unreason in our time is epitomized in the social dynamic which sustains the scientific and technological progress through which the resources of nature are ever more artfully exploited. This triumph of human rationality derives its impetus from the uncontrolled interaction of processes that are rooted in irrational social behavior: the wasteful consumption of the advanced capitalist societies, the fearful military contest between capitalist and socialist blocs, the struggles within and among socialist societies concerning the correct road to the future, and the increasing pressure on third-world nations and their populations to yield fully to economic development and ideological commitment. In the passions that prompt such behavior are forged the ineluctable chains which bind together technology and political domination at present.*

* See appendix—"Technological Rationality: Marcuse and His Critics" —p. 199.

8

THE LIBERATION OF NATURE?

> *It was morality that protected life against despair and the leap into nothing, among men and classes who were violated and oppressed by men: for it is the experience of being powerless against men, not against nature, that generates the most desperate embitterment against existence.*
>
> NIETZSCHE, *The Will to Power*

1. Recurring Ideologies

The idea of the domination of nature is both complex and elusive. In conventional usage, however, the phrases "conquest," "control," and "domination of nature" are purged of their inherent ambiguities, and their meaning appears to be self-evident. Whether an author uses them with approval or reprobation, the terms themselves usually merit little more than a hurried mention, despite the fact that they are supposed to designate something of great significance. As long as this practice is widely followed, both the notion of the domination of nature and its apparent opposite, the liberation of nature, will continue to mystify social reality rather than to clarify it. Domination of nature and its surrogates have become labels for a powerful ideology in modern society, and

167

this process not only affects the understanding of them, but also prejudices the meaning of the liberation of nature. Under present circumstances the latter, instead of becoming a rational concept, must remain only a counterideology.

The word "ideology" is used here to designate a network of ideas which serves in a dual capacity, functioning at one and the same time as a revealing and a concealing conception, as simultaneously explanatory and deceptive. A ruling ideology is a visible indicator of far-ranging social contradictions. It is an effort to represent in conscious form the interests that motivate individuals and groups, but its value is determined by the degree to which such persons understand the full implications of their own desires and of the social dynamic which is set in motion by them. The idea of the "natural rights of man," mentioned in this connection in Chapter One, provides an illustration which may now be better understood as a result of the intervening discussion. On the one hand, the doctrine of natural rights formed the chief rallying-cry for the attack on feudal social relations, and as such it revealed the interests of individuals who were united in various struggles to establish different political, economic, and cultural institutions. On the other hand, this doctrine also helped to conceal a fundamental set of contradictions which were part of the structure of that new society (capitalism) from its beginnings and would remain throughout all the stages of its development down to the present. The free, autonomous individual is the supposed subject of these natural rights alike in the spheres of economy, the state, and intellectual life; but the majority of individuals have always been compelled to surrender their autonomy in the labor-process, while the increasing concentration of economic power gradually integrated the areas of politics and culture into a unified system of control. The legend of the equality of rights and individual freedom, however, together with the illusion of popular choice under the conditions of mass democracy, still veil the reality in which the decisions of the few govern the lives of the many. The contradiction between the abstract universal form of the doctrine—the universal equality of rights—and the concrete particular interests of the minority who rule capitalist society remains unresolved.

Mastery of nature is also an influential contemporary ideology whose revealing and concealing features have been analyzed in the preceding chapters. Needless to say, it is not understood as such in conventional usage; rather, it functions there as a convenient label or an apparently unambiguous sign for a shifting ensemble of social factors. Although its disturbing elements are often intimated—for example, the association between the domination of nature and the domination of man—normally little effort is expended in attempting to delineate them accurately. The dangers involved in perpetuating such shallow formulations must be recognized. In the historical career of an ideology the concealing veil wears thin and a progressively wider gap separates its overt promises from the intractable reality to which they refer. The dichotomy between its universal form and the real particular interests that lie behind it becomes insupportable. If this contradiction is not comprehended and transcended, the ideology itself comes under increasing attack from counterideologies; and most importantly of all, its genuinely progressive aspects may be engulfed in the hatred aroused by the negative conditions with which it becomes associated.

The doctrine of natural rights suffered this fate in part. During the nineteenth century the contradiction between the ideology of equal rights and individuality on the one hand, and the social reality of industrial capitalism on the other, destroyed the original historical force of that ideology. The claim that its promises could be realized within the institutional framework of capitalist society was rejected in theory and practice by the labor movement and radical intellectuals. But in the transition from liberalist to monopoly capitalism, the ideology was reestablished despite the fact that its material foundations (a market economy of individual entrepreneurs without concentrations of economic power) had disappeared forever. Whatever may be the reasons for this occurrence, the widening chasm between promise and reality was (and continues to be) a threat to the possibilities for safeguarding the positive results achieved so far under the banner of natural rights. The danger is that this doctrine becomes *identified* with the particular social system in which it evolved; by virtue of the continuing propagandistic self-advertisements of the "free world," according to which the blessings of individual liberty are to be found exclusively therein, it comes to be regarded by others as *only* an ideology of "bourgeois society." Whereas its negative features can still be concealed domestically, the foreign enemies of the "free world" who experience the destructive force of its antagonism cannot remain unaware of them. The inherited cultural traditions of the non-Western world are only now being adapted to modern social conditions. The fact that by and large these other nations encounter only the negative aspects of capitalist society —economic exploitation and military aggression—while its positive cultural and institutional accomplishments are mostly reserved for the enjoyment of its own citizenry, renders the institutionalization of those positive values elsewhere so much more difficult. In short, the failure to transcend the contradiction between positive and negative aspects within capitalist society is responsible in some measure for the tragic neglect among its opponents of the invaluable legacy contained in the doctrine of individual rights.

Mastery of nature may succumb to a similar fate. Like the notion of natural rights, it was formulated in universal terms as a great *human* task, the benefits of which would accrue to the species as a whole rather than to any particular group. "Relief of the inconveniences of man's estate" was its announced objective. More than three centuries later, however, the goal remains immeasurably distant. The circumstances that thwarted its realization are a matter both of the defects in its conception and also of the specific social dynamic within which it developed; both factors have been explored in the preceding chapters. As long as the original conception retains its efficacy, the contradictory aspects of mastery over nature will remain uncomprehended or at best only dimly perceived. Yet there are abundant indications that it will not continue to retain its efficacy in the coming years; its message of universality no longer elicits the same fervent response, and the paradoxical juxtaposition of the conquest of nature and the conquest of man appears increasingly sinister to many individuals. The attenuation of its universalistic credo—mastery of nature as a general human undertaking—has the effect of causing it to be identified with specific social institutions and tendencies in the immediate environment. These institutions are the organizing centers for the ongoing scientific and technological progress whereby (according to the established orthodoxy) mastery over nature is accomplished: the vast, interlocking, public and private bureaucracies of governments, corporations, military establishments, and university research groups.

These are the same organizations through which social and political domination, domestic and foreign, terroristic and nonterroristic, dictatorial and democratic, is exercised in ever-changing ways. In the concrete social reality mastery of nature characterized as scientific and technological progress increasingly appears to be inseparable from the actual institutional network that plans and directs the successive stages of that activity. Of course the universalistic credo is strenuously maintained both in theoretical analysis and in official pronouncements intended for public consumption. But the belief is harder to sustain in each passing epoch: mastery of nature seems less a grand enterprise of the species than a means of upholding the interests of particular ruling groups. To social theorists and the general public alike, the positive and negative features of progress appear to be indissolubly mixed, thus giving rise to the danger that, were the negative aspects to become intolerable, the entire enterprise would turn into an object of hatred. Since in terms of its usual formulation mastery over nature cannot satisfy the expectations that are associated with it, one day it may fall victim to a counterideology promising a different road to happiness. And under such circumstances its positive as well as its negative aspects may be spurned.

"Liberation of nature" could serve in the future as such a counterideology, in a manner not yet determined. What Horkheimer described under the heading of the "revolt of nature" was an early example of it in a repressive and irrational form. Fascist ideology used the concept of nature in its "blood-and-soil" theories as a weapon against rationalism: the realm of nature was glorified as the original and true source of feeling, inspiration, and action in contrast to the supposedly distorted conceptions arising out of intellectual reflection; a return to this source was alleged to supply a remedy for cultural sickness and a guide for correct behavior. Nature would be liberated from the shackles of civilization. In concrete terms this meant the calculated removal of restrictions on aggressive behavior which was then channeled into specific directions in order to serve the interests of domination; for whereas critical rational thought was ruthlessly suppressed, at the same time the technological rationality embodied in the established institutions of the corporate economy was carefully preserved. The severe social crisis which gave birth to fascism brought to the surface many latent contradictions that were epitomized in the use of antitechnological propaganda as a mask for a regime which exploited modern technology to the fullest in the service of domination.

Under entirely different circumstances the idea of the liberation of nature appeared in an ideological guise in the early stages of nineteenth-century socialist theory. Around 1840 a group of thinkers, among whom Ludwig Feuerbach was the seminal influence, framed their demands for socialism and communism in terms of a return to a "natural" order. The new society would constitute the reversal of a process of etiolation, as it were, for according to them life and happiness are not separated in the realm of nature; they thought that society should conform to nature so that human aspirations might be realized. In a typical passage, found in an article entitled "Cornerstones of Socialism," one of them wrote:

> . . . love, friendship, justice and all the social virtues are based upon the feeling of natural human affinity and unity. Up to the present, these have been termed obligations and have been imposed upon men; but in a society founded upon the consciousness of man's inward nature, i.e., upon reason and not upon external compulsion, they will become free, natural expressions of life. In a society which conforms to nature, i.e., to reason, the conditions of existence must therefore be equal for all its members, i.e., must be general.[1]

Marx and Engels subjected the efforts of this circle, whom they labeled the "true socialists," to a lengthy critique in their joint work, *The German Ideology* (written 1845–46). There they explicitly rejected this naturalistic socialism; but Marx's 1844 manuscripts reveal how close he himself had been to that group's position only a short time before.[2]

Such notions have recurred periodically since that time. The idea of living in harmony with nature, rather than attempting to dominate it, has a certain seductive charm. Indeed as long as it is formulated correctly—not as an apotheosis of primitivism and a proscription of all mechanization, but rather as the elimination of wasteful production and the destruction of the environment—there is much value in it. However, when it is employed in dogmatic fashion merely as a slogan, as a means of expressing generalized displeasure with the prevailing behavior, it tends to become only a counterideology and as such to lose most of its effectiveness as an oppositional force. Certainly this ideology is far less dangerous than its distorted fascist counterpart; but used merely as a slogan it causes the battle lines of social conflict to be incorrectly situated. The potential field of unification among individuals and groups in opposition to the prevailing system is needlessly reduced if people feel that they are faced only with a choice between total acceptance or total rejection of modern technology. The dogmatic mode that characterizes messages of salvation ultimately betrays the possibilities for rationally directed social change.

Nor can serious philosophical thought hope to construct unaided a bridge to a better future. Nothing in the preceding exposition was intended to suggest that a different concept of nature could provide a cure for the problems under discussion there. The prospect of such a solution arises periodically in philosophy and is apparently irrepressible, but it is inevitably frustrated. After an initial period of critical insight Feuerbach succumbed to this illusion, for example, but he was not the last to do so. In modern times the founders of phenomenology have revealed a special penchant for similar undertakings. Max Scheler prophesied the coming of a new elite of metaphysicians as bearers of social progress, and Husserl joined to his own analysis of the mathematization of nature a description of his phenomenology as equivalent to a "religious conversion" that could transform the life of mankind.[3] The error in this expectation is the failure to see in systems of ideas the refracted image of unresolved social contradictions and conflicting possibilities; the mere substitution of different ideas is purely abstract and cannot of itself improve the social situation. These thinkers propose a change of ideas as a mechanic would recommend a change of oil, but in the intellectual realm this can only result in the substitution of one ideology for another, thereby perpetuating the mystification of social reality.

The beginnings of a solution for the dilemmas concerning the conquest of nature that are posed in contemporary thought (as illustrated in Chapter One) are to be found in an account of the *historical function* of the idea of mastery over nature as a fundamental ideology of modern society. The preceding analysis was designed to lay the foundations for such an account, which will be presented in Section 2 below. A few preliminary remarks will indicate the general perspective in which it is framed.

In Part One of this book I have attempted to describe the process whereby the domination of nature came to be identified with scientific and technological progress. Subsequently I have tried to show why this identification is illicit and to discuss what qualifications must be attached to this conventional judgment. Thus in characterizing the domination of nature as an ideology I am not thereby ascribing to science itself an ideological function. In fact the concept of science *does* play an ideological role in modern society. That is a separate problem which has not been treated in this study, however; only insofar as science is conceived as a crucial element in the domination of nature does it share the fate of the latter as an ideology. Science itself becomes ideological when a particular method of arriving at scientific knowledge succeeds in establishing a claim to be the *only* valid entry into the entire realm of objective understanding. This is also a crucial issue in contemporary intellectual history; but while it is peripherally related to the problem under discussion here, it represents within the same social framework a different constellation of historical tendencies, the investigation of which is a distinctive subject of its own.

Every significant ideology undergoes changes in its historical function over time. Therefore any critique which attempts only to specify its truth value as seen from the perspective of the present—that is, to abstract a fixed content or meaning from the historical flux of its fortunes—would seriously misrepresent its significance as a motivational force in human practice. On the one hand, the idea of mastery over nature helps us to understand what common hopes bind together the present structure of human action and its preceding stages; on the other, the failure to recognize its changing concrete role in action during different periods distorts the past as well as the present. In fact it is precisely the extraordinary success of this idea in motivating social action that is largely responsible for the successive shifts in its historical function: as a critical factor in historical development it cannot and does not remain immune to the dynamic which it helped to set in motion. The reciprocal interplay of idea and reality, theory and practice, determines the changing physiognomy of both.

To recall the analogy with the doctrine of natural rights once again: during the seventeenth and eighteenth centuries that notion served to unite individuals and groups in assaults on outmoded and repressive institutions; this constituted its specific ideological function in that period. Quite apart from this circumstance there is its enduring positive contribution to civilization, namely, the establishment of new institutions in which individual liberties were partially protected, together with the promise of further development toward their full realization. The negative side of progress in this respect was its mystification of the real economic relationships among social classes. In the twentieth century its ideological function is quite different, for those older precapitalist institutions have been vanquished. Particular improvements in broadening the scope of individual rights have been made, to be sure; but the now-official ideology conceals the permanent threat of internal catastrophe represented in the recurring fascist and neofascist movements which may destroy all the victories won in the preceding centuries. And as indicated earlier, the ideology struggles against its own internal contradictions and mystifies the global social dynamic by picturing the world as a battleground in which the preservation of individual freedom depends wholly upon the "security" of advanced capitalist society. Here too the success of the doctrine contributed to the alteration of the general historical situation, and thus in part to the transformation of its own function.

Beginning in the seventeenth century the idea of the mastery of nature spurred an attack upon outmoded scientific and philosophical dogmas and helped to initiate a qualitative change both in the understanding of nature and in the possibilities for the satisfaction of human needs: this was *its* specific ideological function at that time. The lasting positive aspect of its service was (as formulated so well by Bacon) to break the tyrannical hold of despair over the consciousness of human technical possibilities and to encourage the conviction that men could fundamentally alter the material conditions of existence. Its negative dimensions—so well disguised in Bacon's *New Atlantis*—were its exclusive focus on modern science and technology as the designated instruments for the mastery of nature and its ability to mask the connection between their development on the one hand, and the persistence of social conflict and political domination on the other.

The change in its ideological function during the course of the twentieth century is a direct result of the fact that the new modes of thought championed by it triumphed completely over the older dogmas. The notion of mastering nature through scientific and technological progress, at first but the frail progeny of a few innovating thinkers, has become a semiofficial ideology whose traces can be found even in popular literature and the daily newspaper. The negative dimensions mentioned above were only latent possibilities in the seventeenth century, but in the intervening period they have become real barriers to social progress. The concealed ambiguities analyzed above in Chapters Five through Seven have turned a once-creative idea into a rhetorical slogan that is repeated endlessly in a mechanical fashion. The only remedy for social ills is said to be the ever more competent scientific and technological mastery of nature; but to some extent the medicine feeds the disease.

Once creative and progressive ideologies, natural rights and mastery over nature have been transformed into sterile, mystifying dogmas. In neither case was this a necessary fate, nor is it yet too late to refashion them and render them enlightening concepts once again. But this is possible only if their positive and negative features are clearly distinguished and if they are freed from their self-defeating attachment to earlier stages of social development. The doctrine of individual freedom must surrender its archaic vision of universal participation in a market economy without state intervention, composed of producers with equal economic weight, and the idea of the domination of nature must yield up its fond dream of human technological power over nature that remains socially and politically innocent. Until such time the past will rule the present with a despotic hand, distorting men's understanding of their own activity as a carnival room of mirrors distorts an image.

2. Society and Nature

Throughout this book I have described the various ways in which philosophical and sociological studies have conceived the idea of the domination of nature and its relation to modern science. They may be summarized briefly as follows: (1) The most influential schools of modern philosophy, including the rationalist tradition from Descartes to Fichte as well as empiricism and positivism, developed the epistemological and ontological grounds for viewing scientific thought as the basis for the human mastery over nature. (2) As opposed to all preceding types of science, modern science as inaugurated by Galileo and others has a fundamentally instrumentalist conceptual structure and thus an *a priori* technological character. (3) The domination of nature is linked both logically and historically with capitalist or bourgeois society. (4) Modern science is related logically and historically to bourgeois society. These ideas and the interrelationships among them have been developed by both Marxist and non-Marxist thinkers in a series of critical studies that go beyond the superficial categories of conventional usage and attempt to probe more deeply into the relationship between the changing structure of social relations, on the one hand, and changing attitudes toward nature on the other.

It has not been my purpose in this book to analyze any of these theories as such; an adequate historical and critical study of them, which has not yet been done, could be the subject of four separate accounts. I have tried only to indicate in passing some of their strengths and weaknesses insofar as they touch upon the approach adopted here for understanding the notion of mastery over nature. In my view these theories have not succeeded—either by themselves or taken together—in offering an accurate statement of the historical function of mastery over nature as a crucial ideology in modern society. To be sure all of them possess some truth, and all have contributed significantly to answering the questions posed at the beginning of this study. The work of many hands is to be found on the canvas exhibited here, and I hope to harmonize these earlier efforts with a few additional broad strokes so that for the first time there will emerge a clear picture of the problem we have been discussing.

This unifying thesis may be formulated thus: the vision of the human domination of nature becomes a fundamental ideology in a social system (or of a phase in the development of human society considered as a whole) which consciously undertakes a radical break with the past, which strenuously seeks to demolish all "naturalistic" modes of thought and behavior, and which sets for itself as a primary task the development of productive forces for the satisfaction of human material wants.

The first social system in the history of civilization in which these tendencies are found is Western capitalism. They pertain to its essential structure as a unique mode of social relationships (to the "spirit of capitalism," in Max Weber's terminology). They were at the outset only latent possibilities championed against hostile institutions and ideologies by weak, scattered groups who were themselves unaware of the potential consequences of their demands; their dynamic was unleashed with the gradual triumph of capitalist society over its predecessor. Of course they do not represent the only basic characteristics of capitalism, for they are intrinsically related to other developments, such as the commodity form of production. And at no time have these tendencies ever been fully recognized and comprehended by the competing social classes: both the cunning of reason and the cunning of unreason served to conceal from individuals and groups the full implications of their activity.

Yet although they originate with capitalist society, these trends are not confined exclusively to it. They reappear under socialism to the extent to which this social system undertakes to realize the unfulfilled and unfulfillable promises of capitalism. As the means of industrialization and economic development first in the Soviet Union and then elsewhere, socialism directs the construction of the material basis necessary for the emergence of individuality and general freedom. Only after this is achieved, and only after the fateful global confrontation of the competing systems is overcome, will socialist society reach the point at which the nineteenth-century radical thinkers assumed it would begin its course. In their present phase, however, the socialist societies necessarily share the vision described above, and the conquest of nature is reincarnated there as a familiar ideology.[4]

Precapitalist societies share almost universally a common feature, namely, a reliance on various "naturalistic" categories as a basis for social organization, distinctions of rank, the allocation of work, maintenance of political domination, and so forth.[5] In other words, the principles which provide a justification for the allocation of roles and power are grounded upon the assertation that they conform to the "order of nature." Nature's regime is assumed to be eternal and unalterable, and therefore resistance to change is the primary goal of social authority. Moreover, the natural is equivalent to the good, so that any deviation from the established conditions could only bring disaster for the society as a whole. The transmission of rank, function, and wealth through successive generations in accordance with family or kinship associations —those "natural" or biological relationships which supply the foundations for social organization—is the usual way in which conformity to the supposed order of nature is perpetuated over long periods. Of course there are many variations of this basic scheme that are related to different types of cultural formations and to different levels of complexity in social relations; but common to them all is the fact that the concept of nature has a *prescriptive force* for human ethical and political consciousness. The idea that the order of nature furnishes the certain standard for the order of society was one of the earliest manifestations of a functioning ideology.

At the origins of conceptual thought the human spirit projects itself into nature: the voice of nature rules human conduct through nonhuman beings (demons and so forth), physical locations (oracles), and other media. But in the development of civilization nature is gradually "despiritualized" —this is one of the primary accomplishments of religion— and begins to lose its prescriptive force. This is a slow process, however, which changes qualitatively only with the coming of capitalism.

Western feudal society still depended heavily on naturalistic categories to provide a rationale for the system of estates and its related values. The principle that qualitative levels and distinctions are enshrined in nature was valid equally for physical science and social theory. The identification of the natural with the good was preserved at the highest level of abstraction in the scholastic absorption of Aristotelian philosophy. Continuity with the past, maintained through ennobled families which were closely identified with specific territorial domains, was a paramount concern of social policy. The Judaeo-Christian religious tradition with its sharp separation of spirit and nature is profoundly antinaturalistic in inspiration; but by itself it could not eliminate all naturalistic categories, since in feudalism the material relations of production, based on the absolute distinction between lord and serf, were rooted in them.

During the period of transition from feudalism to capitalism much was transmitted from the one to the other. Modes of thought are exceptionally resilient and tend to persist long after their supporting framework has evaporated; thus naturalistic categories may be detected still today, despite the fact that the general tendency of capitalist society is to eliminate them. For example, the notion of "obedience" to the "natural laws" of a free-market economy has been represented not as reflecting solely the dictates of prudence and the calculus of self-interest, but rather as possessing far loftier ethical overtones. In times of economic crisis this residual naturalism inhibited business and political leaders from "interfering" with the supposedly unalterable laws of the market: its principles were thought to be ordained by nature rather than by men, and men believed that to violate them was to court social disaster. Only the severe breakdown during the Great Depression effectively destroyed this archaic naturalism and prepared the way for the widespread acceptance of a managed capitalist economy in which market mechanisms are assiduously manipulated through the offices of government.*

---

* In modern social theory the ideas of natural rights and the laws of nature are not necessarily naturalistic categories, since "nature" is often identified with "reason"—for example, in the writings of Hobbes and Locke.

Capitalism undermined all the social foundations upon which naturalistic modes of behavior were based. Its fundamental tenet is the abstract equality of individuals. All distinctions of role, function, and rank established according to birth, lineage, or any other natural condition are held to be improper. The anomaly of slavery, for example, could be tolerated only as the enslavement of a different race and was never recognized as legitimate in social theory even under these conditions: when in a desperate maneuver the Southern theorists in the United States recalled Aristotle's argument that some men are "slaves by nature," they were bitterly attacked for it. The notion of a complete equality among all individuals, together with the idea of the opposition between nature and society, are the cornerstones of the social-contract theory, which was itself the great intellectual weapon used against the defenders of the old society. Locke's *Two Treatises of Government* are a perfect illustration, for in the first he attacks Robert Filmer's naturalistic political theory before offering in the second the social-contract theory as a better way of understanding human society.

Hobbes showed the most penetrating insight into the significance of the new principles of societal organization. He boldly dismissed the received Aristotelian dictum that social membership is "natural" for men: a proclivity for such association, he said, hardly enables us to distinguish men from bees or ants. Society is not natural; on the contrary, men are driven to it under the compulsions of terror and the fear of death. Society is the great "artificial body" through which they seek to preserve themselves from the self-destructive promptings of their own nature—human nature.[6] Nature provides no model for the structure of social relations: in the later formulation of the theory by Rousseau, men are said to develop a "second nature" within society. Hobbes also expressly denies that any natural conditions such as differences of strength can form the basis for social distinctions, since the superior cunning of the weaker man often overcomes the physical prowess of the stronger.

The divorcing of society from "nature" loosened an entire series of restrictions on social interaction and paved the way for an enormous expansion of productive forces. Marx described capitalism as a permanently revolutionary socioeconomic system.[7] Liberated from a host of traditional restraints, economic activity (in its specific form as commodity production) penetrated far deeper than ever before into all spheres of social life and continually opened up new dimensions of productive activity. Although all such steps are necessarily contained within this specific form of production, the growing supremacy of economic concerns over all others creates both the reality and the illusion of perpetual social change. The doctrine of continuous progress became widely accepted; the past appeared as burdened with irrational fetters, and the idea of a better future dominated the present. "Development" became a fundamental category of social thought.

The characteristics of this process are well known, and only a few details will be mentioned here for illustrative purposes. The working day was lengthened, and the working year was extended by the elimination of the frequent festivals and holidays connected with agricultural life; child and female labor were far more intensively exploited. Labor was transformed into work, the latter being distinguished from the former by virtue of its complete secularization as the last vestiges of its sacred attachments disappeared. The businessman replaced the warrior and the cleric as the guiding figure in social policy. The role of the family, the church, and regional cultural traditions in the transmission of values was weakened; popular consciousness and behavior became both more uniform and more fluid. Finally, the system expanded aggressively into the noncapitalist world and subjected traditional cultures and societies to intense pressures, ideological as well as material, which only the oldest and strongest could resist for any length of time.

This is the historical context in which mastery over nature evolved as a fundamental social ideology. The final "despiritualization" of nature, the steady weakening of the prescriptive force carried by the concept of nature, created the ethical and ideational void that was subsequently filled by the newer conception. Without a doubt Francis Bacon evinced the subtlest understanding of the intellectual dynamic involved here. He argued that the dominant philosophy of his time unwittingly projected false notions into nature, or in other words, that human concerns were being read into the order of nature in a wholly abstract manner. He proposed to replace this immediate identification of human interest with the natural order— which according to him must inevitably be frustrated—by a mediated one: once liberated from the immediate interest of locating in nature the source of particular ethical and social standards, human consciousness would be free to develop a universal picture of nature's own *modus operandi,* which could then be applied subsequently in the service of human desires. Restraint of immediate interests was the mediating device that would make possible the real fulfillment of human interests. In effect he showed that "obedience to nature" was a practical or technical injunction rather than a support for moral imperatives.

The elimination of naturalistic categories means in this context that the concept of nature ceases to be a basis for limitations on the scope of human behavior. Since it is no longer burdened by an attachment to particular immediate interests, nature appears increasingly under the aspect of universality as the *generalized object* of investigation, experimentation, and an open-ended technological applicability. When human consciousness no longer projects itself into external nature in search of security and validation for standards of conduct, nature can be viewed merely as a system of matter in motion, as purely an object or field of conquest for human theoretical and practical intelligence. The experience of being dominated *by* nature—that is, by external standards grounded in nature—gives way to the expectation of achieving domination *over* nature.

In fact there was a kind of "natural" transition from the one to the other, a continuum of meaning that helped to determine the specific, enduring form in which the idea of human domination over nature was cast. Domination by nature was always a matter of concrete political domination, for specific ruling groups (for example the feudal nobility) were the ones who attempted to legitimate their authority by means of naturalistic categories. Thus domination by nature was experienced in the context of politically administered force and power. The ever-present, concrete political context influenced the later representation of domination over nature, in which (as I argued earlier in analyzing Bacon's terminology) a suspect political imagery has prevailed down to the present. Domination by nature sustained the power of particular social strata; domination over nature has conventionally been conceived as the winning of power for the species as a whole. The fundamental error committed by Bacon and his followers was to characterize the latter by analogy with the former. For there is a qualitative difference between the senses of "power" in the two cases, as I shall try to demonstrate presently, and the failure to recognize it has been responsible in large measure for perpetuating the hidden contradictions in the predominant conception of mastery over nature.

The same historical setting which we have been discussing also explains why mastery over internal nature (or human nature) is a correlate of mastery over external nature. The decay of the naturalistically grounded bonds that imprisoned persons within defined ranks and functions opened broad new dimensions for the exercise of individual initiative and the pursuit of self-interest. At the same time, however, the liberated energies had to be channeled in certain determinate directions lest the social fabric be destroyed by "irrational" drives. The market economy provided the basic mechanism.

Competition and aggression found ever-widening outlets as the relative social significance of economic activity increased, and simultaneously the legitimated aggressive drives were contained and checked within a socially defined framework. When individuals are no longer considered as belonging "by nature" to a recognized social rank, society must assume complete responsibility (through laws or rules promulgated by its ruling strata) for controlling behavior. This process has its own dialectic, of course: as the obvious creations of men the societal rules are far more vulnerable to criticism and transformation than were the naturalistic categories.

The growing influence of the idea of human domination over nature in modern society, therefore, is a reflection of these three historical tendencies: a decisive break with the past, an elimination of naturalistic categories, and a qualitative change in the possibilities for the satisfaction of needs. Its historical function as an ideology was to contribute to the achievement of those objectives, and it did so by altering the ways in which men evaluated the significance of their own theoretical and practical activity. The degree to which it succeeded in this endeavor is truly astounding.

When its positive work is clearly recognized, the corresponding limitations of the idea also become apparent. Only in relation to the specific historical context in which the modern version of this idea arose and developed is it possible to perceive the most general element of inadequacy in the prevailing conception of mastery over nature. Its essential limitations were present at the outset, but they have become manifest only in recent times, precisely because this ideology has been so successful in changing deeply rooted attitudes. The fault lies in the fact that a fundamentally *static* image has been employed to represent an intensely dynamic historical process.

The Baconian formulation of the idea of human domination over nature, which became the *leitmotif* of subsequent thought down to the present, is internally consistent only in a *religious* context. Bacon's own writings preserve this context, as I attempted to show in Chapter Three; but in laying the foundations for the secularization of this idea, Bacon introduced certain contradictions which he, like his followers, failed to notice. The source of those contradictions is that domination of nature, conceived as the possession of power *over* nature by the human species as a whole, is an idea which makes sense only in relation to the absolute separation of spirit (God) and nature in Judaeo-Christian theology—and thus is an idea which cannot be secularized without losing its internal harmony.

In this theology, spirit dominates nature as the creator of nature. Man shares the privilege of domination inasmuch as he is the only natural being which participates in the realm of spirit. The fact of domination is an eternal condition, that is, it is without any reference to time or change, for time is created by spirit simultaneously with nature. In Christianity the great moment of the irruption of spirit into nature—the incarnation of Christ—occurs as a means of restoring the original foundation of God's work through the redemption of man from the consequences of the Fall. Domination over nature is *a priori:* man's portion of it is the gift of God rather than his own accomplishment. And this is related to another *a priori* condition, namely, the unity of the species which inherits Adam's prerogatives. Domination over nature as a religious notion pertains to "man as such," not to particular men in their desperate search for means of satisfying their mundane needs.

Considered as aspects of the secular course of human history, on the other hand, quite obviously neither domination over nature nor the unity of the species may be viewed as *a priori* conditions; rather they are (or are expected to be) the outcome of stages in the actual development of civilization. The internal connection between the two within a theological setting breaks down during their transfer to a nonreligious context, where each serves to distort the conception of the other. Domination over nature is wrongly represented as an achievement that will bind together a bitterly divided species; conversely, the abstract idea of man (in the phrase "man's conquest of nature") hides the fact that the actual agents in this process are individuals and societies in violent conflict among themselves. The secular attempt to achieve power over nature by means of science and technology takes place in the historical context outlined above, that is, in a social setting transformed by accelerating change and a dynamic of intersecting forces that defies men's efforts to bring it under rational control. But the ideological representation of that attempt—the concept of the domination of nature—fails in its intention to fully comprehend the reality which stands before it. Too static and brittle, the concept is stretched and finally broken in the social flux: contemporary theory reluctantly concludes that the positive and negative features of technological progress are inseparable, that somehow domination over nature has been firmly shackled to its apparent opposite, domination over men.

This cannot be regarded simply as a matter of theoretical confusion. The fixed order represented in the static image was supposed to *contain* the dynamic reality, not just in theory, but in actual practice as well. Bacon understood this perfectly, as is shown in a passage quoted earlier: "Only let the human race recover that right over nature which belongs to it by divine bequest, and let power be given it; the exercise thereof will be governed by sound reason and true religion." [8] Unfortunately he was mistaken. Neither reason nor religion was capable of guiding the search for power over nature and of preventing it from becoming self-destructive. Contrary to his great expectations, modern scientific and technological rationality failed to escape the hold of the more powerful mechanisms kept in motion by irrational social conditions. The prevailing concept of the domination of nature must share some of the responsibility for this occurrence, for it has helped to conceal from the consciousness of generations the fact that their existing social institutions cannot contain the destructive potentialities of their scientific and technological competence.

Is it still possible to rectify this mistake? The present situation offers some hope that the answer may be in the affirmative. In many of the sources discussed in Chapter One, a representative sample drawn from contemporary literature which illustrates what I have called the "prevailing conception" of the idea under discussion here, the paradoxical character of mastery over nature has been acknowledged, albeit rather timidly. Moreover, the growing revulsion against the irrational uses of scientific and technological capabilities— such as flights to the moon and products which poison the environment—in the most favored nations may mean that it is not too late to apply those capabilities toward extinguishing the global social conflagrations in which all peoples, space conquerors and cow worshippers alike, will otherwise be engulfed. A new understanding of the idea of the mastery of nature, in suggesting a different perspective for evaluating the kinds of activity through which mastery of nature might be realized, may make a small contribution to that endeavor. Such a perspective must enable us to recognize the immense positive consequences of the Baconian doctrine as well as the necessity of transcending the hollow formulas to which it has been reduced at present. A few suggestions along these lines are offered below by way of a return to the questions posed at the beginning of this study.

3. Nature and Human Nature

The static image of human domination over nature that was derived from religion is inadequate in a secular context because it presents the subject or agent (man) as actively commanding an object (nature) *with neither subject nor object undergoing any essential change as a result of this activity.* From the perspective of religion, man's nature is defined as part of the original foundation of the world, and the labor of time is devoted to the task of restoration. Readmission to the state of grace is a spiritual drama played against a passive background (the realm of nature) which itself remains unaffected, except in the quasiheretical notion of the "redemption of nature" that periodically returns in religious mysticism. The purposes of religion may indeed be well served by this conception, but the secular variant of it falsifies the character of the ongoing relationship between man and nature. In human history a reciprocal dialectic is at work: "By thus acting on the external world and changing it, [man] at the same time changes his own nature." [9] But the conventional image obscures the second element indicated in Marx's perceptive statement. When Bacon envisaged a revolutionary science and technology that would "shake nature in its foundations," he failed to understand that at the same time the related conditions of human behavior would be transformed no less dramatically.

The static representation of the agent was responsible for fostering the illusion that the activity through which the mastery of nature would be pursued was itself "mastered," that is, under rational control from the outset. This illusion is exhibited most clearly in *New Atlantis,* where Bacon presents his philosopher-scientists as making independent decisions concerning what discoveries and devices would be revealed to the social authorities. What is presupposed in this picture is that the directors of the scientific and technological enterprise possess both the requisite abilities and the political power to exercise and enforce judgments not merely in technical matters, but more importantly with respect to the *social* consequences of their achievements. But even at the most advanced stages of technological development, when the social system appears to be so dependent upon the steady progress of this enterprise, its personnel are unable (and perhaps also unwilling) to assert any independent sway over the social applications of their efforts. The wartime agonies of certain atomic physicists constituted a futile outburst which for a brief moment called attention to the normal state of affairs, namely, a silent acquiescence in the social impotence of scientists that is shared both by the general public and by the practitioners of that trade.

In Bacon's conception there is a presumption of an *a priori* link between the increasing control of external nature and a complementary element of self-control in human behavior which would guide the social applications of the newly gained powers. The connection is literally embodied in the managers of Solomon's House. Yet he never explained adequately the bases (psychological or otherwise) upon which this presumption was founded, nor, obviously, was he struck by the anomaly in the larger situation: whereas the conduct of the isolated scientific elite is bettered by virtue of their scientific activity, that of the citizenry from whom they are physically and spiritually divorced—but in whose service they labor—apparently remains unimproved. The Enlightenment sought to overcome this paradox by maintaining that scientific rationality would diffuse through all spheres of culture and would eradicate the sources of irrational behavior in society as a whole. This expectation too failed to materialize, for reasons that were mentioned earlier.

These original errors in the Baconian vision were incorporated in the ideology of mastery over nature that came to play such a crucial role in modern thought, but this circumstance did not prevent it from being a positive force over a period of several centuries. I wish to emphasize the point once more: in referring to this conception of mastery over nature as an ideology I do not mean to suggest that from the beginning it was a "false" or obscurantist doctrine. On the contrary, it represented a bold attempt to anticipate a set of important historical tendencies; the popularity which it has enjoyed is convincing evidence of its efficacy as an explanatory notion, and it continues to portray truthfully a significant part of the social reality in which we live. To transcend this vision means not to reject it outright, but rather to preserve its positive elements within the outlines of a new formulation that will be more adequate in the context of the present.

The idea of the mastery of nature must be reinterpreted in such a way that its principal focus is *ethical or moral development* rather than scientific and technological innovation. In this perspective progress in the mastery of nature will be at the same time progress in the liberation of nature. The latter, like the former, is a rational idea, a concept, an achievement of human thought; therefore the reversal or transformation which is intended in the transition from mastery to liberation concerns the gradual self-understanding and self-disciplining of *human* nature. As a rational idea "liberation" can apply only to the work of consciousness, to human consciousness as an aspect of nature, and not to "nature" as a totality. The task of mastering nature ought to be understood as a matter of bringing under control the irrational and destructive aspects of human desires. Success in this endeavor would be the liberation of nature—that is, the liberation of human nature: a human species free to enjoy in peace the fruits of its productive intelligence.

Again a caveat: ethical progress and scientific-technological progress are not simple opposites. What is of value in each depends in some measure on the accomplishments of the other. The development of scientific rationality, as I suggested earlier, is one crucial presupposition of any ethical advance, in that it counteracts the human propensity to project irrational structures into external nature and to be tyrannized by those projections. Bacon and the Enlightenment thinkers understood this well, and their recognition of this problem is a lasting monument of human culture. However, from the perspective of the times in which they lived it was impossible to foresee the full importance of the complementary proposition: ethical progress as a general phenomenon affecting all individuals is a fundamental presupposition of scientific and technological innovation, for in the absence of the former the latter tends to become self-destructive. Some of the seventeenth- and eighteenth-century theorists knew this also, but their remarks on this point have been largely ignored. Now the necessary balance between the two must be restored.

Science and technology will be no less important for us when they are deprived of their rank as the principal forces in the mastery of nature. In fact, I think that their great accomplishments would be far more secure as a result. As we have seen, according to the prevailing conception the conquest of nature is regarded as the enlargement of human power over nature, with science and technology serving as the instruments of this drive, for the purpose of satisfying material wants. Pursued in this manner, mastery over nature inevitably turns into mastery over men and the intensification of social conflict. A vicious circle results, imprisoning science and technology in a fateful dialectic of increasing mastery and increasing conflict. The attractive promises of mastery over nature—social peace and material abundance for all—remain unfulfilled. The real danger that the resulting frustration may be turned against the instruments of mastery themselves (science and technology) must not be underestimated. As integral factors in the ascending spiral of domination over internal and external nature, they are bound to an irrational dynamic which may destroy the fruits of their own civilizing rationality.

The liberation of science and technology from the thrall of this dynamic is a task that primarily involves the reconstruction of social institutions. Theoretical reflection plays a part in this process by suggesting changes in the categories through which individuals and groups interpret the significance of their activity. To dethrone science and technology as the guiding forces in the mastery of nature is a necessary step not only for them, but for the idea as well. Freed from a historical association that has become anachronistic, the idea of the mastery of nature would be open to new meanings. Considered as moral progress, it would indicate more forcefully the most demanding challenge that confronts us: not to conquer external nature, the moon, and outer space, but to develop the ability—widely dispersed among all individuals in society—to use responsibly the technical means presently available for the enhancement of life, together with an institutional framework which will nourish and preserve that ability.[10]

In a magnificent short essay, C. S. Lewis exposed all of the most fundamental flaws in the conventional wisdom on this subject. "In what sense," he asked, "is Man the possessor of increasing power over Nature?" He suggested that when we consider any of the technological instruments by means of which this power is supposedly exercised, we find that in the concrete contexts of everyday life the individual has access to such power only under certain determinate circumstances. These limiting conditions are defined by property relationships and the distribution of social authority. Moreover, as often as not the individual is the "patient" rather than the agent of this power, since it is used against his interests and desires when these conflict with the wishes of others. He concludes: "From this point of view, what we call Man's power over Nature turns out to be a power exercised by some men over other men with Nature as its instrument."

The finest insight in Lewis's argument is his insistence that the concept of power over nature must be seen in the perspective of the relationships among successive generations of the human species. Power over nature is only one of the ways in which each generation seeks to win control over its destiny, and in so doing, it inevitably confronts the past and predetermines the future:

> In order to understand fully what Man's power over Nature, and therefore the power of some men over other men, really means, we must picture the race extended in time from the date of its emergence to that of its extinction. Each generation exercises power over its successors: and each, in so far as it modifies the environment bequeathed to it and rebels against tradition, resists and limits the power of its predecessors. This modifies the picture which is sometimes painted of a progressive emancipation from tradition and a progressive control of natural processes resulting in a continual increase of human power. In reality, of course, if any one age really attains, by eugenics and scientific education, the power to make its descendants what it pleases, all men who live after it are the patients of that power. They are weaker, not stronger: for though we may have put wonderful machines in their hands we have pre-ordained how they are to use them. . . . There is therefore no question of a power vested in the race as a whole steadily growing as long as the race survives. The last men, far from being the heirs of power, will be of all men most subject to the dead hand of the great planners and conditioners and will themselves exercise least power upon the future.[11]

Lewis saw clearly that the continual emphasis on the winning of power over nature tended to block the awareness of a crucial dilemma: How can we insure that this power will be used responsibly? And when all social problems are converted into technical problems, what can serve as rational bases for criteria of choice? His discussion brings to life an issue that has remained unresolved since the seventeenth century. I pointed out earlier that in Bacon's view religion was to guarantee the responsible use of power over nature and to serve as the existential "ground" for science as a social enterprise. Descartes shared this belief, as we have seen: Christian morality was to supply the guiding beacon for the individual's life-course while the new scientific methodology was revolutionizing the social order. But religion failed to justify these expectations, and Lewis realized that the problem had become acute in the twentieth century. Yet he insisted that a set of traditional values, anchored in religion and impervious to the corrosive force of technological rationality, remains the only secure buttress of civilization.

The diagnosis is accurate, but the proposed remedy seems far too ineffectual. Western religion has lost its hold over the domain of practical activity, and the increasingly secularized character of social behavior renders unlikely the prospect that it may someday restore its hegemony. However, a theologically based interpretation of the relationship between intelligence (spirit) and nature continues to govern the concept of human domination over nature. The present secular context requires a very different interpretation, namely, one in which mastery of nature is understood as an advanced stage in human consciousness wherein intelligence is able to regulate its relationship to nature (internal and external) in such a way as to minimize the self-destructive aspects of human desires. Unlike the static religious image, this conception presupposes an interplay between intelligence and nature in which both change fundamentally and continuously. The subject-object (man) can overcome the irrational promptings of its own nature; and nature considered as an object of human desire, when it ceases to be regarded primarily as a source of power (and is thus condemned to feed the fires of human conflict), can become instead the wellspring of happiness.

The secular foundations of the mastery of nature in this new sense would be a set of social institutions in which responsibility and authority are distributed widely among the citizenry and in which all individuals are encouraged to develop their critical faculties. Habermas and others have begun to explore in detail the range of present possibilities for the emergence of such institutions. For the work of mastering nature cannot be confined to a minority if we understand that objective as a matter of moral progress. If this process fails to improve the condition of the majority, sooner or later the traditional causes of irrational behavior, running their course in the midst of steadily enlarged destructive capacities, will bring about the end of civilization. Bensalem's dazed onlookers and magnificent potentates alike must transform themselves radically if the labors of Solomon's House are to be preserved. To control their scientific and technological ingenuity men must first cease to be astonished by it and to request blessings which it is incapable of bestowing on them.

At the end of a little book first published in 1928, Walter Benjamin remarked that we should not regard the essence of human technique as the ability to dominate nature. Rather, he suggested, we should view it as the mastery of the relationship between nature and humanity.[12] Such an attitude would do justice to the subtle interplay of internal and external nature. Mastery of the relationship between nature and humanity, a kind of mastery no longer bound to repressive demands arising out of the structure of domination in society, could bring to fruition the progressive hopes embodied in the original notion of the domination of nature.

APPENDIX

Technological Rationality: Marcuse and His Critics

The immediate relationship between technology and domination, forged by the struggle for the satisfaction of needs, marks all human technology with an intrinsic political character. "Techniques" comprise not only tools but equally as importantly the organization and training of human labor: Lewis Mumford illustrated this point well in his argument that the first great machine in history consisted of the forced-labor gangs that built the Egyptian pyramids, together with the state administration which planned and supervised their work.[18] The purposeful organization and combination of productive techniques, directed either by public or private authorities, has been called "technological rationality." Depending upon the level of cultural development, it is normally linked with a specific type of scientific rationality, that is, with a more abstract understanding of nature's physical processes.

Nineteenth-century socialist thinkers emphasized the significance of technological development for the cause of human liberation, specifically with reference to the construction of an adequate material basis for satisfying needs through a minimum of labor. Few of them—least of all Marx and Engels —could be called technological determinists, however, since they also stressed their expectation that a complementary improvement in the exercise of rational judgment among the majority of the population would establish the proper conditions for the enjoyment of freedom. Marx called this the

Footnotes to the Appendix will be found in the notes to Chapter Seven.

of the "general intellect." This expectation has so
ppointed, and the development of a consciousness
mong the majority in the industrially advanced
been blocked, while technological expertise has
ceeded from success to success.

An analysis of the social and social-psychological grounds
of false consciousness has been a major problem for the critical theory of society (Horkheimer, Adorno, Marcuse, and
others), and the steady growth of technological rationality
throughout the twentieth century was recognized by them as
a crucial factor in this problem. But although brief comments
on it may be found at many points in their writings, no systematic treatment of this concept appeared before Marcuse's
exposition in Chapter Six of *One-Dimensional Man,* entitled
"From Negative to Positive Thinking: Technological Rationality and the Logic of Domination." What he attempted to
argue there was clearly stated: "It is my purpose to demonstrate the *internal* instrumentalist character of [modern] scientific rationality by virtue of which it is *a priori* technology,
and the *a priori* of a *specific* technology—namely, technology
as form of social control and domination." [19] Marcuse presents technological rationality as only one of the basic forms
of "one-dimensionality," but he repeatedly stresses its all-
embracing character in advanced industrial society. Jeremy
Shapiro, who has further developed this notion, has defined
one-dimensionality as "political domination masked as technical rationality." [20]

This chapter of Marcuse's work elicited a storm of criticism. Rolf Ahlers and Hans-Dieter Bahr, for example, have
found in it a final confirmation of Marcuse's attachment to
the errors of Heideggerian philosophy; Habermas, on the
other hand, sees it as the revelation of the long-hidden motif
of Romantic *Naturphilosophie* in his thought, the idea of a
"new" science and technology cleansed of the stain of domination. [21] Others have tried to document his deviation from the
orthodoxy of Marx and Engels on this point. [22] In my view all

of these criticisms are more or less irrelevant because they focus on a few isolated quotations and fail to consider the main lines of Marcuse's argument. Far superior are the essays by Claus Offe and Joachim Bergmann (published in the volume *Antworten auf Herbert Marcuse*) which specifically deal with the concept of technological rationality. Offe details the obscurities that result from Marcuse's use of the terms "industrial society" and "technological society" and correctly demands a far more precise characterization of the actual social interests which determine the structure of advanced capitalist society. Bergmann contends that Marcuse's notion of technological rationality attempts unsuccessfully to subsume diverse mechanisms of social domination (economic rationalization in Max Weber's sense, the politics of mass democracy, and scientific-technological progress) under a single category; the frequent reference to a shadowy "apparatus" of control inhibits the empirical study of how these different mechanisms actually function at present.

The deficiencies noted by Offe and Bergmann arise both out of Marcuse's general manner of exposition and also out of the specific faults of his argument in *One-Dimensional Man*. Like his original colleagues in the "Frankfurt School," he tends to couch his thought in an epigrammatic style. Much reflection on complex issues is compressed into brief passages which, when examined in isolation, sometimes appear inconsistent; and the reasoning as a whole lacks smooth transitions from one idea to the next. Thus one can find many comments, not all of which are harmonious, on technology and the domination of nature scattered throughout his publications over a period of more than thirty years. However, a careful study of them reveals a consistent approach to this problem which offers guidance for the interpretation of individual passages, especially those found in *One-Dimensional Man*. In the following pages I will try to present and evaluate the consistent features of Marcuse's thought on the subject of the present study.

The concept of technological rationality first appears prominently in an article written by Marcuse in 1941, entitled "Some Social Implications of Modern Technology." There it was contrasted with two other forms of reason, namely, individual rationality and critical rationality. Individual rationality according to Marcuse was the hallmark of bourgeois society in its initial phases of development; new social forces demanded freedom for the individual to exercise his reason in all spheres of activity, material (economic) as well as intellectual. This struggle was carried out against the established social interests, so that individual rationality was at the same time critical rationality, opposition to outmoded social institutions and ideologies. The ideas of liberalism were built upon the premise that individuals were or could be autonomous, that is, persons whose thought and decisions were the result of critical self-reflection and self-interest.

The promises of a liberal society were thwarted. With the further development of capitalist society "the process of commodity production undermined the economic basis on which individualistic rationality was built," and this rationality was "transformed into technological rationality." [23] In other words, the economic units of production continually expanded in size, until the laws of the free market were effectively abolished and small individual entrepreneurs lived at the mercy of the great corporations which direct the economy. This transformation in the sphere of production destroyed the material foundations of individualistic rationality; henceforth productive rationality was embodied in large organizations, and the individual had to adjust himself and conform to the dictates of the "rationality of the apparatus." But this organized or technological rationality offers the individual little scope for critical reflection:

> Individuals are stripped of their individuality, not by external compulsion, but by the very rationality under which they live. . . . The point is that today the apparatus to which the

individual is to adjust and adapt himself is so rational that individual protest and liberation appear not only as hopeless but as utterly irrational. The system of life created by modern industry is one of the highest expediency, convenience and efficiency. Reason, once defined in these terms, becomes equivalent to an activity which perpetuates this world. Rational behavior becomes identical with a matter-of-factness which teaches reasonable submissiveness and thus guarantees getting along in the prevailing order.[24]

Yet there is no simple opposition or contradiction between critical and technological rationality. Both constantly change their content, partially in response to each other. The former seeks to preserve the substance of individual rationality—the idea of autonomous individuals who are capable of organizing their lives under the conditions of freedom—in a time when social changes have destroyed the earlier promises of liberal society. It seeks to identify those tendencies within technological rationality (for example, the democratization of functions) which might still establish a basis for individual autonomy and freedom under changed social circumstances. The emphasis throughout Marcuse's analysis here is on the restrictive social forms by which technological rationality is forced to serve the interests of domination rather than freedom: "Technics hamper individual devlopment only insofar as they are tied to a social apparatus which perpetuates scarcity, and this same apparatus has released forces which may shatter the special historical form in which technics is utilized."[25] Marcuse points out that the antitechnological attitude is used as an ideology (for example, in fascism) to disguise the powerful alliance between the advanced rational technologies of production and terroristic political domination. The fault lies not with technological rationality itself, but with the repressive social institutions which exploit the achievements of that rationality to preserve unjust relationships among men. A sentence from the later work summarizes perfectly the theme of this article: "In the social reality, despite all change, the domination of man by man is still the historical continuum that links pre-technological and technological Reason." [26]

The internalization of mechanized work routines, patterns of consumption, and socially dictated leisure activities are some of the principal means by which individuals surrender their critical faculties to the requirements of the production system. In their experienced needs and desires individuals reproduce the necessities of the institutions which oppress them. Heteronomy, the unreflective internalization of behavioral norms, impedes the possible formation of autonomous judgments among the majority, and thus

> the technology and technics applied in the economic process are more than ever before instruments of social and political control. The satisfaction of needs (material and intellectual) takes place through scientific organization of work, scientific management, and the scientific imposition of attitudes and behavior patterns which operate beyond and outside the work process and precondition the individuals in accord with the dominant social interests.[27]

The highly original insight found in both Horkheimer's and Marcuse's work is an insistence on a necessary connection between mastery of external nature and mastery of internal nature. The enormous social enterprise through which is undertaken the conquest of external nature by scientific and technological rationality demands a complementary control of internal nature, including the discipline of the work process and the expression of need and satisfaction. This insight prompted their continuing interest in a critical social psychology and especially in a radical interpretation of Freud's metapsychology of civilization.

In *Eros and Civilization* Marcuse outlined a theory of the relationship between the development of the ego and the conquest of external nature very similar to Horkheimer's (as presented above).[28] But it is in one of his separate essays on Freud that he states most clearly his conception of the development of domination in civilization:

As soon as civilized society establishes itself the repressive transformation of the instincts becomes the psychological basis of a *threefold domination:* first, domination over one's self, over one's own nature, over the sensual drives that want only pleasure and gratification; second, domination of the labor achieved by such disciplined and controlled individuals; and third, domination of outward nature, science, and technology.[29]

This multiple domination, by which the individual is subjected to requirements imposed upon him from without, is not the eternal opposite of freedom, but rather its presupposition. A "threefold freedom"—moral, political, and intellectual— emerges from the work of domination. Having rationally mastered their own inclinations and having constructed the material basis for the satisfaction of needs, individuals may utilize the inheritance of civilization for the enjoyment of freedom.

The legacy of domination does not disappear of its own accord, however. One of the central themes in *Eros and Civilization* is an analysis of the process whereby domination is perpetuated under social conditions which contain the actual grounds for the realization of individual freedom. "Surplus repression," representing the elements of domination which do not serve the interests of maintaining civilization, increases as the promise of liberation dawns. A decisive break with the "continuum of domination," a change from quantitative to qualitative progress, becomes an essential task of revolution. *Soviet Marxism,* dating from the same period, is in part an attempt to explain why the first great socialist revolution was unable to accomplish this break. Judging Stalinism in much the same way as does Isaac Deutscher, Marcuse emphasizes the superhuman effort that was required in order to transform a "backward" society in the face of antagonism and mortal danger from without, as a means of explaining how the liberating message of Marxism became an ideology that justified oppression.

The theme of *Soviet Marxism* required Marcuse to deal

once more with the function of technology in social change, and he did so in a way that is consistent with his earlier article on the social implications of modern technology. He maintains that it is the repressive use of technology "which makes for its dehumanizing and destructive features: a restrictive social need determines technical progress." From this viewpoint he concludes that "the truly liberating effects of technology are not implied in technological progress per se; they presuppose social change, involving the basic economic institutions and relationships." [30] But already in the new preface to this book, written for the publication of the paperback edition in 1960, there is a different phrasing of the problem which foreshadows the treatment of it in *One-Dimensional Man*. In reply to criticisms Marcuse insisted that he wished to retain his "emphasis on the all-embracing political character of the machine process in advanced industrial society." This statement is curious, because it does not really characterize the argument of *Soviet Marxism* at all, but rather anticipates the theme of the later book which was undoubtedly in the early stages of preparation at that time.

The best critiques of *One-Dimensional Man* have pointed out that its major flaw lies in its use of the concepts of "advanced industrial society" and "technological society." In various places Marcuse claims to be exploring the converging tendencies of highly developed societies, both capitalist and socialist, but the actual content of the book—its specific topics, examples, and illustrations—is drawn exclusively from intellectual trends in Western capitalist society. One could not even say that the analogous tendencies in socialist society had been discussed previously in *Soviet Marxism,* since the two works are not really complementary. The idea of a "technological universe," a comprehensive and powerful mode of activity which "shapes the entire universe of discourse and action, intellectual and material culture"—an idea that is forcefully outlined in the preface to *One-Dimensional Man*—

is never clarified sufficiently. This vague but menacing notion recalls both the language and the philosophical standpoint of Ellul's *The Technological Society* (a work that is never mentioned by Marcuse), despite the fact that one would not otherwise suspect any affinity between these two quite dissimilar thinkers.

The basic unclarity in the notion of a technological universe is carried over into the discussion of technological rationality in Chapter Six. There Marcuse attempts to unite two different propositions and in so doing creates the serious confusions that have aroused the ire of his critics. In a passage quoted earlier he describes the twofold intention of his approach: (1) to show that modern scientific rationality is inherently instrumentalist; (2) to demonstrate that this instrumentalist rationality is the impetus behind "a *specific* technology— namely, technology as form of social control and domination." Together they are "the realization of a specific historical *project*—namely, the experience, transformation, and organization of nature as the mere stuff of domination." In this endeavor "science, *by virtue of its own method* and concepts, has projected and promoted a universe in which the domination of nature has remained linked to the domination of man . . ." [31] The difficulty in his exposition is his failure to show clearly the interrelationship between science as instrumentalist rationality and a technology that reinforces political domination.

In my view Marcuse succeeds in demonstrating his first point. Scheler earlier advanced a similar contention, as I indicated above in Chapter Five; Marcuse's discussion complements Scheler's and also adds many important ideas. The treatment of this point constitutes the real substance of that chapter in *One-Dimensional Man,* and the second point is not really defended at all. Several objections to the latter may be raised: first, it does not follow from the initial argument; second, if taken literally it contradicts what Marcuse has said elsewhere many times; and third, as stated above it gives the erroneous impression that what is required is a new technology entirely different from the technology of domination.

The form of rationality that characterizes the modern natural sciences is instrumentalist in a specific sense, but this does not mean that it is inherently bound to technology as an instrument of domination. What constitutes the historical connection between scientific rationality and the progress of domination over nature and men is a specific constellation of *social* forces, as I have tried to show earlier. That same scientific rationality can become—in a different social setting—a force for the self-mastery of human nature, without altering its substance in any way. But Marcuse seems to demand a change in the form of scientific rationality itself when he says that as part of a qualitative turn in the direction of progress "science would arrive at essentially different concepts of nature and establish essentially different facts." [32] His failure clearly to explain this statement has led to the charge that despite specific disavowals on his part he is indeed calling for a revival of the Romantic philosophy of nature. Only in a later essay did he state in a straightforward manner his belief that

> there is no possibility of a reversal of scientific progress, no possibility of a return to the golden age of "qualitative" science. . . . The transformation of science is imaginable only in a transformed environment; a new science would require a new climate wherein new experiments and projects would be suggested to the intellect by new social needs. . . . Instead of the further conquest of nature, the restoration of nature; instead of the moon, the earth; instead of the occupation of outer space, the creation of inner space . . . [33]

The present science is a science whose tasks and problems are determined in a social setting of conflicts, wars, and perpetual ideological mobilization; the new science would be guided by the goals of peace, happiness, and the beautification of the environment in an ongoing process of rational discourse and interaction among scientists and nonscientists. The progress of its specific internal rationality would be affected only insofar as these new goals would produce different priorities in the allocation of resources for research and experimentation.

The existing connection between scientific rationality and political domination is to be found in the "absolutization" of a particular scientific method as the *only* valid source of objective knowledge. This is Horkheimer's point, and I think that Marcuse's work as a whole is in accord with it. Earlier, in Chapter Six, I indicated that it was this absolutization that produced a crisis in modern ethics which has not yet fully run its course; in commenting on problems of ethics Marcuse has made similar remarks. The fact that this absolutization becomes a significant social phenomenon—for example, the *predominance* of a behavioral methodology in the social sciences out of which arises refined techniques for the control and manipulation of human actions—can be explained only with reference to a particular constellation of social interests, and *not* with reference to the instrumentalist character of scientific methodology, either in the natural sciences or derivatively in the social sciences. To be sure, such instrumentalism provides the *a priori* basis for better control of external and internal nature; but as Marcuse himself explained so concisely in his essay on Freud, such control is also the precondition of freedom under changed social conditions.

Certain inconsistencies in Marcuse's work simply cannot be eliminated. In my opinion the careless formulations in *One-Dimensional Man* contradict the consistent features of his thought as found both prior and subsequent to that book. Among later writings his important essay on Max Weber provides the best illustration: there he returns to the basic outlook of his first article on technology while providing a deeper insight into the historical relationship between rationality and domination in modern society. He maintains that Weber identifies "technical reason with bourgeois capitalist reason," and that technological rationality is thus at the same time also "capitalist rationality." Weber's concept of fate describes "a society in which the law of domination appears as objective technological law"; to the extent that the first phenomenon is not recognized in the second, the concept of technical reason becomes an ideological mask:

> . . . *this* technical reason reproduces enslavement. Subordination to technology becomes subordination to domination as such; formal technical rationality turns into material political rationality (or is it the other way around, inasmuch as technical reason was from the beginning the control of "free" labor by private enterprise?).[34]

From the outset technical reason was political reason, and vice versa; the first has always been limited and distorted in the interests of domination. Thus it has essentially different potentialities in relation to different social systems: "Capitalism, no matter how mathematized and 'scientific,' remains the mathematized, technological domination of men; and socialism, no matter how scientific and technological, is the construction or demolition of domination."[35] Technological rationality can aid the perpetuation of domination under socialism, but socialism retains the promise of destroying domination; capitalism remains inevitably bound to a structure of domination.

Marcuse's article closes with the idea that within a different social framework technical reason "can become the technique of liberation." This point is reinforced in the strongest possible terms in *An Essay on Liberation:*

> Is it still necessary to state that not technology, not technique, not the machine are the engines of repression, but the presence, in them, of the masters who determine their number, their life span, their power, their place in life, and the need for them? Is it still necessary to repeat that science and technology are the great vehicles of liberation, and that it is only their use and restriction in the repressive society which makes them into vehicles of domination?[36]

Another essay written at about the same time notes that technological progress can contribute to a fateful continuity between capitalism and socialism unless deliberate steps are taken to counteract such a development. To break the link between technology and domination under socialism means not to repeal technological progress itself, but to "reconstruct the technical apparatus in accordance with the needs of free men, guided by their own consciousness and sensibility, by their autonomy." [37]

Moreover, Marcuse has recently returned to an idea first stated at the end of the new preface to *Soviet Marxism,* namely, that different cultural traditions in the non-Western world may aid the presently "underdeveloped" nations in avoiding the repressive and destructive uses of advanced technologies and in constructing a modern technological capability that is firmly related to rational needs at every stage.[38] Certainly this is no more than a chance, one which is continually threatened by the intense military-economic and ideological pressures on the third world emanating from the "developed" areas. Should such a program succeed even in the slightest degree, however, there would arise the possibility of a counterinfluence that might affect the ongoing efforts in the industrially advanced societies to shatter the bonds between political domination and technological rationality.

The consistent features in Marcuse's thought on this subject may now be summarized as follows: (1) the continuum of domination in the social relations among men shapes the way in which technological rationality develops—and in part the latter determines the evolution of the former; (2) scientific and technological progress in themselves do not undermine the social foundations of domination—on the contrary, the "technological veil" can serve to support them; (3) scientific and technological rationality constitute one of the essential preconditions for freedom, and in a liberated society they are among the indispensable requisites for the enjoyment of freedom. The apparent logical inconsistency between the second and third points is rather a *real historical contradiction.* There is an ongoing dialectic of rationality and irrationality in society which magnifies simultaneously the possibilities for intensified domination on the one hand, and for liberation on the other. The full force of that dialectic is not necessarily broken in the transition from capitalism to socialism; it must become the specific objective of individuals associated in the struggle for liberation to accomplish that rupture.

Marcuse's general conception of the mastery of nature functions as a concise resumé of the complex issues outlined in the preceding paragraph. The conflicting and partially contradictory elements in the mastery of nature are concealed in the usual formulations of that idea, as I have tried to show earlier. But Marcuse makes a fundamental distinction which illuminates those contradictions:

> Pacification [of the struggle for existence] presupposes mastery of Nature, which is and remains the object opposed to the developing subject. But there are two kinds of mastery: a repressive and a liberating one. The latter involves the reduction of misery, violence, and cruelty. . . . All joy and all happiness derive from the ability to transcend Nature—a transcendence in which the mastery of Nature is itself subordinated to liberation and pacification of existence.[39]

Liberation is equivalent to the nonrepressive mastery of nature, that is, a mastery that is guided by human needs that have been formulated by associated individuals in an atmosphere of rationality, freedom, and autonomy. Otherwise mastery of nature might—and does—serve to perpetuate and intensify domination and irrationality. What is essential is to articulate the specific objectives of mastery over nature in relation to human *freedom* rather than to human *power*. The conventional interpretation of this notion emphasizes the latter and neglects the former, and in so doing succumbs to the cunning of unreason. For the pursuit of ever greater power over nature on the social plane, within the framework of repressive institutional structures, solidifies the existing relations of domination and weakens to a corresponding degree the ability of individuals to shape their destiny through autonomous interaction.

NOTES AND REFERENCES

1. The Cunning of Unreason

1. *Essays, and Wisdom of the Ancients:* "Daedalus, or Mechanical Skill"; "Icarus and Scylla and Charybdis, or the Middle Way"; "Sphinx, or Science."
2. *Daedalus, or Science and the Future,* pp. 83, 85.
3. *Icarus, or the Future of Science,* pp. 7, 12.
4. *Ibid.,* pp. 40–42.
5. See Barrington Moore's essay, "The Society Nobody Wants."
6. For a discussion of this point see Northrop Frye, "Varieties of Literary Utopias."
7. *Utopia and its Enemies,* p. 108.
8. *Ibid.,* p. 3.
9. "Final Report" of the Intergovernmental Conference of Experts on the Scientific Basis for Rational Use and Conservation of the Resources of the Biosphere, p. 15.
10. "Marx est-il dépassé?" in *Marx and Contemporary Scientific Thought,* p. 296.
11. "Commentary," in *The Technological Order,* ed. C. Stover, p. 254.
12. *The Dreams of Reason,* pp. 16, 57.
13. "Utopias and the Living Landscape," in *Utopias and Utopian Thought,* ed. F. Manuel, p. 137.
14. *Walden Two,* pp. 76, 112, 126.
15. *Philosophy of Democratic Government,* pp. 291–292.
16. Robert Boguslaw, *The New Utopians,* p. 204.
17. Nell Eurich, *Science in Utopia,* pp. 270–271. See generally Floyd Matson, *The Broken Image,* Part One.
18. Guy Gresford, "Qualitative and Quantitative Living Space Requirements," in "Final Report," *op. cit.,* Annex IV, p. 1.

213

19. The best introduction to this much-discussed subject is an essay by Lewis Mumford, "Utopia, the City and the Machine."
20. E. F. Murphy, *Governing Nature*, pp. 11–12.
21. Kenneth Keniston in *The New York Times*, 6 January 1969, p. 143.
22. *Reason Awake*, p. 178.
23. *Lectures on the Philosophy of History*, p. 34.
24. Marx Wartofsky, *Conceptual Foundations of Scientific Thought*, p. 29.
25. *One-Dimensional Man*, p. 240.
26. Georg Lukács, *History and Class Consciousness*, pp. 234 ff.
27. *One-Dimensional Man*, p. 236.

2. Mythical, Religious, and Philosophical Roots

1. *The Forge and the Crucible*, pp. 99–100.
2. *Ibid.*, p. 75.
3. "The Historical Roots of our Ecologic Crisis," in his *Machina ex Deo: Essays in the Dynamism of Western Culture;* the quotation is from p. 93.
4. *Ibid.*, p. 86. In his recent book, *The Dominion of Man*, John Black has also presented a valuable discussion of the theological background to the contemporary concern with ecology.
5. Alexandre Kojève, "L'Origine Chrétienne de la science moderne."
6. David Asselin, S. J., "The Notion of Dominion in Genesis 1–3," pp. 281n, 283.
7. Clarence Glacken, *Traces on the Rhodian Shore*, pp. 310, 349.
8. St. Thomas Aquinas, *Summa Theologica*, Part I, Quest. 3, Art. 1, and Quest. 96, Art. 2, in the translation by the Fathers of the English Dominican Province. Cf. Glacken, pp. 295–302 and *passim*.
9. Jaroslav Pelikan, "Cosmos and Creation: Science and Theology in Reformation Thought," p. 468.
10. *The Human Condition*, p. 139n. Cf. Glacken, p. 166: "But one must not read these [Biblical] passages with modern spectacles, which is easy to do in an age like ours when 'man's control over nature' is a phrase that comes as easily as a morning greeting."
11. Both quoted in Glacken, pp. 296, 481.
12. See H. W. Janson, *Apes and Ape-Lore in the Middle Ages and the Renaissance*, Ch. 10, especially pp. 305 ff.
13. Quoted in Glacken, p. 463.

14. Frances Yates, *Giordano Bruno and the Hermetic Tradition,* pp. 60–61.

15. This paragraph is based upon D. P. Walker, *Spiritual and Demonic Magic from Ficino to Campanella.*

16. Paolo Rossi, *Francis Bacon: From Magic to Science,* Ch. 2; the quotation from Agrippa is on pp. 18–19.

17. See Yates, Chs. 1–3 and E. M. Butler, *The Myth of the Magus.*

18. *Oration on the Dignity of Man,* in *The Renaissance Philosophy of Man,* ed. E. Cassirer *et. al.,* pp. 248–249; Yates, pp. 107 ff.

19. Yates, pp. 148 ff.; Cassirer, "Mathematical mysticism and mathematical science."

20. Yates, p. 156; see also pp. 447 ff. Max Scheler advanced a similar theory in the 1920's; see below, Chapter Five.

21. Lynn Thorndike, *A History of Magic and Experimental Science,* Vol. VII: *The Seventeenth Century,* pp. 8–9.

22. Quoted in Rossi, pp. 19–20.

23. E. M. Butler, *The Fortunes of Faust.*

24. Quoted in Yates, p. 136.

25. Eliade, *The Forge and the Crucible,* pp. 150 ff.; Jung, *Psychology and Alchemy (Collected Works,* Vol. XII), pp. 228 ff.

26. Quoted in Frank Manuel, *A Portrait of Isaac Newton,* p. 173. Cf. Eliade, p. 169: ". . . alchemy prolongs and consummates a very old dream of *homo faber:* collaboration in the perfecting of matter while at the same time securing perfection for himself."

27. Jung, *Alchemical Studies (Collected Works,* Vol. XIII), pp. 127–128; Eliade, pp. 172–174. Cf. C. S. Lewis, *The Abolition of Man,* pp. 52–53.

3. Francis Bacon

1. This section is based primarily on the following studies: Paolo Rossi, *Francis Bacon: From Magic to Science,* and *Philosophy, Technology and the Arts in the Early Modern Era;* the two books on Bacon's philosophy by F. H. Anderson; Ernst Cassirer, *Das Erkenntnisproblem,* Vol. II, Ch. 1; Margery Purver, *The Royal Society: Concept and Creation;* and Benjamin Farrington, ed., *The Philosophy of Francis Bacon,* Part One.

2. *Sämtliche Werke,* Vol. III, p. 65.

3. *The New Organon,* in *The Works of Francis Bacon,* Vol. IV, pp. 247–248.

4. Quoted in F. H. Anderson, *The Philosophy of Francis Bacon,* pp. 14–15.

5. *Thoughts and Conclusions,* translated in *The Philosophy of Francis Bacon,* B. Farrington, ed., p. 92.
6. *The New Organon: Works,* IV, 115.
7. *Valerius Terminus of the Interpretation of Nature,* III, 222.
8. Preface to *The Great Instauration,* IV, 20.
9. See Anderson, p. 153, and the references given there.
10. *Valerius Terminus: Works,* III, 222.
11. *Preparative towards a Natural and Experimental History,* IV, 263.
12. *The Advancement of Learning* (translation of *De Augmentis Scientiarum*), IV, 287.
13. *The New Organon,* IV, 108.
14. *The Masculine Birth of Time,* in Farrington, p. 62.
15. *Of the Proficience and Advancement of Learning, Divine and Humane: Works,* III, 283, 302.
16. *The Great Instauration,* IV, 24.
17. *The Advancement of Learning,* IV, 343.
18. *Description of the Intellectual Globe,* V, 506; *Thoughts and Conclusions,* in Farrington, p. 93.
19. *The Advancement of Learning: Works,* IV, 296.
20. *The Great Instauration,* IV, 29; *Description of the Intellectual Globe,* V, 505–506.
21. On the sexual imagery in alchemy see Gaston Bachelard, *La formation de l'esprit scientifique,* Ch. 10.
22. *Civilization and its Discontents,* p. 101 (translation slightly changed).
23. *Life against Death,* p. 102; see also pp. 236, 316.
24. The impact of Bacon's *New Atlantis* is discussed in Nell Eurich's *Science in Utopia* and Frank Manuel's *The Prophets of Paris.*
25. The best modern edition of *Utopia* is in volume IV of More's *Complete Works,* published by Yale University Press; *New Atlantis* will be found in volume III of the standard edition of Bacon's *Works.* There are also many separate editions of each. For commentaries see Martin Schwonke's *Vom Staatsroman zur Science Fiction* and especially the article by Robert P. Adams, "The Social Responsibilities of Science in *Utopia, New Atlantis* and After."
26. *Complete Works,* IV, 135.
27. Quoted from the edition of G. C. Moore Smith, pp. 32–33.
28. "Bacon's Man of Science," in *The Rise of Science in Relation to Society,* ed. L. Marsak, p. 50.
29. A. B. Gough, introduction to his edition of *New Atlantis,* p. xxx.
30. *New Atlantis,* ed. G. C. Moore Smith, p. 45.

31. *Thoughts and Conclusions*, in Farrington, pp. 92–93; cf. *The New Organon: Works*, IV, 114.

4. *The Seventeenth Century and After*

1. Lynn Thorndike, *A History of Magic and Experimental Science*, Vol. VII: *The Seventeenth Century;* Glacken, *Traces on the Rhodian Shore*, pp. 482 ff.
2. Peter Gay, *The Enlightenment: An Interpretation*, Vol. II: *The Science of Freedom*, p. 160.
3. Cited in Thorndike, p. 544; see generally Ch. 19.
4. Robert Lenoble, *Mersenne ou la naissance du mécanisme*, and Thorndike, Ch. 14.
5. Paolo Rossi, *Philosophy, Technology and the Arts in the Early Modern Era*, pp. 30–31; see also the articles of Edgar Zilsel on this point.
6. *Ibid.*, p. 42.
7. *Histoire du renouvellement de l'Académie Royale des Sciences*, in *Oeuvres diverses*, Vol. III, pp. 6 ff.; see Leonard Marsak, *Bernard de Fontenelle: The Idea of Science in the French Enlightenment*.
8. Gilles-Gaston Granger, *La mathématique sociale du Marquis de Condorcet;* Thomas L. Hankins, *Jean d'Alembert: Science and the Enlightenment;* Isabel F. Knight, *The Geometric Spirit: The Abbé de Condillac and the French Enlightenment*.
9. *Recueil des éloges historiques lus dans les séances publiques de l'Institute Royal de France*, Vol. I, p. 30.
10. Quoted in Margery Purver, *The Royal Society: Concept and Creation*, pp. 95–96.
11. *The Philosophical Works of Descartes*, Vol. I, p. 119.
12. *Capital*, Vol. I, p. 390, footnote (translation changed).
13. Richard Kennington, "René Descartes," in *History of Political Philosophy*, eds. L. Strauss and J. Cropsey, p. 391.
14. *Doctrine Saint-Simonienne: Exposition*, p. 463; see also pp. 338 and 467.
15. See Alfred Schmidt's excellent book, *Der Bergiff der Natur in der Lehre von Marx* (forthcoming in English translation under the title *The Concept of Nature in Marx*), and also the sections on Marx in Jürgen Habermas, *Knowledge and Human Interests*. Kostas Axelos's version of his thought, *Marx, penseur de la technique: De l'aliénation de l'homme à la conquête du monde*, attempts (unsuccessfully) to present Marxism as an extreme form of Saint-Simonianism; a useful corrective is Jean Fallot, *Marx et la machinisme*.

16. *Capital*, Vol. I, p. 177.
17. *Grundrisse der Kritik der politischen Ökonomie*, p. 593; translation quoted from Herbert Marcuse, *One-Dimensional Man*, p. 36.
18. *Capital*, Vol. III, p. 861.
19. *Capital*, III, 800; *Anti-Dühring*, p. 392 (translations changed).
20. *The World View of Physics*, p. 199, and *The History of Nature*, p. 71.
21. Randall, *Aristotle*, pp. 2–3; Heisenberg, *Physics and Philosophy*, pp. 196–197.
22. *The Physicist's Conception of Nature*, p. 24.
23. Ernst Cassirer, *Idee und Gestalt*, p. 69.
24. *Goethe und der deutsche Idealismus*, p. 7.
25. *Der Übergang vom feudalen zum bürgerlichen Weltbild*, pp. 6, 12.
26. See further Alexandre Koyré, *From the Closed World to the Infinite Universe*, Ch. 4.
27. The two quotations are from Borkenau, pp. 54 and 347.
28. Grossman, "Die gesellschaftlichen Grundlagen der mechanistischen Philosophie und die Manufaktur," p. 192.
29. See David Joravsky, *Soviet Marxism and Natural Science, 1917–1932*, and L. R. Graham, *The Soviet Academy of Sciences and the Communist Party, 1927–1932*.
30. For a bibliography on this topic consult T. K. Rabb's article, "Puritanism and the Rise of Experimental Science in England," and Lewis Feuer, *The Scientific Intellectual*, Ch. 1.
31. *Philosophy, Technology and the Arts in the Early Modern Era*, p. 28.
32. "Galileo and Plato" (1943), in *Metaphysics and Measurement: Essays in Scientific Revolution*, p. 16; he refers to Borkenau's book in a footnote on p. 17. See also his *Études Galiléennes*, Vol. I, pp. 6–7; *Études d'histoire de la pensée scientifique*, pp. 60–61; and *Newtonian Studies*, pp. 5–7.
33. *The Social Function of Science*, p. 415.
34. "The Science of Science: A Programme and a Plea," pp. 490–492.
35. *Ibid.*, p. 496.

5. Science and Domination

1. For general accounts of his thought see Maurice Dupuy, *La philosophie de Max Scheler*, and J. R. Staude, *Max Scheler*.

2. *Major Problems in Contemporary European Philosophy*, pp. 150–151.

3. See especially the essay "Nietzsches Wort 'Gott ist tot' " in *Holzwege*.

4. *Die Wissensformen und die Gesellschaft*, p. 68. On this typology cf. Max Horkheimer, "Zum Problem der Wahrheit," in *Kritische Theorie*, ed. A. Schmidt, Vol. I, pp. 259-260.

5. *The Will to Power*, ed. W. Kaufmann, Sect. 480; cf. *Wissensformen*, p. 224. Heidegger has correctly stressed the point that Nietzsche does not use the terms "will" and "power" in their ordinary senses. See Landgrebe's discussion and references, *Major Problems in Contemporary European Philosophy*, pp. 156–159, and Habermas, *Knowledge and Human Interests*, Ch. 12.

6. *Wissensformen*, p. 94.

7. *Ibid.*, p. 130. Cf. Edgar Zilsel's article, "The Genesis of the Concept of Physical Law."

8. *Ibid.*, p. 122, footnote 2.

9. *Ibid.*, p. 67; cf. p. 277. On this point see also John Dewey, *Experience and Nature*, pp. 115–116.

10. *Wissensformen*, p. 128, and "Phänomenologie und Erkenntnislehre," in *Schriften aus dem Nachlass*, Vol. I, pp. 425–427.

11. *Wissensformen*, p. 125; see also pp. 93–94 and 194.

12. *Ibid.*, p. 198.

13. *Ibid.*, pp. 124 ff.

14. "Über die positivistische Geschichtsphilosophie des Wissens (Dreistadiengesetz)," in *Schriften zur Soziologie und Weltanschauungslehre*, p. 27. Cf. Edgar Zilsel's article, "The Sociological Roots of Science."

15. Scheler, *Wissensformen*, p. 128; Horkheimer, *loc. cit.*

16. See Barrington Moore's essay, "Why we fear peasants in revolt."

17. See further Chapter Six.

18. Jürgen Habermas, "Technical Progress and the Social Life-World," in *Toward a Rational Society*, p. 52.

19. The details of this process are set forth in Chapter Seven.

6. Science and Nature

1. *The Crisis of European Sciences and Transcendental Phenomenology*, p. 66. Also p. 271: "The result of the consistent development of the exact sciences in the modern period was a true revolution in the technical control of nature."

2. *Ibid.*, p. 6.
3. *Ideen II*, pp. 2, 27; cf. *Crisis*, p. 130.
4. *Die Krisis der europäischen Wissenschaften und die transzendentale Phänomenologie*, p. 356 (passage from a section not rendered in the published English translation).
5. *Crisis*, p. 23.
6. *Crisis*, p. 295.
7. *Philosophical Works of Descartes*, Vol. I, p. 95.
8. Lukács was the first to explore this problem in depth: see *History and Class Consciousness*, pp. 121 ff.
9. *Major Problems in Contemporary European Philosophy*, p. 99.
10. Cf. Jürgen Habermas, *Knowledge and Human Interests*, Appendix.
11. *Preparative towards a Natural and Experimental History*, in *Works*, Vol. IV, p. 257.
12. Husserl, *Krisis*, pp. 51–52; the English translation (*Crisis*, p. 51) renders it as "garb of ideas."
13. Cf. C. B. Macpherson, *The Political Theory of Possessive Individualism*, Ch. 2.
14. *Crisis*, pp. 130–131. On this point see also Herbert Marcuse's article, "On Science and Phenomenology."
15. The defoliation campaign is described in the report prepared for the United States Department of Defense by the Midwest Research Institute, "Assessment of Ecological Effects of Extensive or Repeated Use of Herbicides," and is analyzed more critically in the writings of the biologist Arthur Galston. For a description of the proposal to orbit space reflectors see *The New York Times*, 29 December 1966 (p. 10) and 26 May 1967 (p. 4); the plan was apparently vetoed by higher authorities in the government.

7. *Technology and Domination*

1. Max Horkheimer, "Soziologie und Philosophie," in Horkheimer and Adorno, *Sociologica II: Reden und Vorträge*.
2. *Eclipse of Reason*, p. 177.
3. *Ibid.*, p. 176.
4. *Dialektik der Aufklärung* (with Theodor W. Adorno), pp. 20, 27–29.
5. *Eclipse of Reason*, Ch. 1, and "Zum Begriff der Vernunft," in *Sociologica II*.
6. *Dialektik der Aufklärung*, pp. 9 ff.; a brief outline is given in *Eclipse of Reason*, pp. 17–18.
7. *Dialektik der Aufklärung*, pp. 31 ff., 189–191.

8. *Eclipse of Reason*, p. 108. In one of his essays Marcuse cited the following passage from Fichte's *Staatslehre* (1813): "The real station, the honor and worth of the human being, and quite particularly of man in his morally natural existence, consists without doubt in his capacity as original progenitor to produce out of himself new men, new commanders of nature: beyond his earthly existence and for all eternity to establish new masters of nature." Quoted in "On Hedonism," in *Negations*, p. 186.

9. *Eclipse of Reason*, p. 97.

10. *Ibid.*, p. 109.

11. "Zum Begriff des Menschen," in *Zur Kritik der instrumentellen Vernunft*, ed. A. Schmidt, p. 198.

12. The "mere" deployment of such weaponry is regarded as a form of employment (for example, in the 1962 Cuban missile crisis).

13. See Irving L. Horowitz, ed., *The Rise and Fall of Project Camelot* (on Latin America); Eric Wolf and Joseph Jorgensen, "Anthropology on the Warpath in Thailand"; David Ransom, "The Berkeley Mafia and the Indonesian Massacre"; and many other sources.

14. *Toward a Rational Society*, p. 61.

15. See generally *Eclipse of Reason*, Ch. 3, especially pp. 109 ff.

16. *Ibid.*, p. 94.

17. See the UNESCO report referred to in Chapter One.

18. *The Myth of the Machine*, Ch. 9.

19. *One-Dimensional Man*, pp. 157–158 (author's italics).

20. "One-Dimensionality: The Universal Semiotic of Technological Experience," in *Critical Interruptions*, ed. P. Breines, p. 181.

21. Ahlers, "Is Technology intrinsically Repressive?"; Bahr, *Kritik der "Politischen Technologie,"* pp. 56–64; Habermas, *Toward a Rational Society*, pp. 81–90.

22. For example, Peter Sidgwick, "Natural Science and Human Theory," in *Socialist Register 1966*, eds. Miliband and Saville, pp. 182 ff.

23. "Some Social Implications of Modern Technology," pp. 416–417.

24. *Ibid.*, p. 421.

25. *Ibid.*, pp. 423, 429, 436.

26. *One-Dimensional Man*, p. 144.

27. Preface (1960) to *Soviet Marxism*, p. xii.

28. *Eros and Civilization*, pp. 109 ff. Cf. Russell Jacoby's superb brief exposition in his essay "Reversals and Lost Meanings," in *Critical Interruptions*, ed. P. Breines, pp. 66–70.

29. "Freedom and Freud's Theory of Instincts," in *Five Lectures*, p. 12 (author's italics).

30. *Soviet Marxism*, p. 241.
31. *One-Dimensional Man*, pp. 158, xvi, 166 (author's italics).
32. *Ibid.*, p. 167.
33. "The Responsibility of Science," in *The Responsibility of Power*, eds. L. Krieger and F. Stern, pp. 442–443.
34. "Industrialism and Capitalism in the Work of Max Weber," in *Negations*, pp. 206, 214, 222-223 (author's italics).
35. *Ibid.*, p. 215.
36. *An Essay on Liberation*, p. 12.
37. "Re-examination of the Concept of Revolution," in *Marx and Contemporary Scientific Thought*, p. 481.
38. *Soviet Marxism*, p. xvi; "Political Preface 1966" to *Eros and Civilization*, pp. xvii–xx; "The Obsolescence of Marxism?" in *Marx and the Western World*, ed. N. Lobkowicz, p. 415.
39. *One-Dimensional Man*, pp. 236–237.

8. The Liberation of Nature

1. Quoted by Marx and Engels in *The German Ideology*, p. 527.
2. See Alfred Schmidt, *Der Begriff der Natur in der Lehre von Marx*, pp. 109–114.
3. Scheler, *Die Wissensformen und die Gesellschaft*, p. 380; Husserl, *The Crisis of European Sciences*, p. 137. For a brief but incisive comment on this point see Horkheimer, *Eclipse of Reason*, pp. 163–165.
4. For example, see the article "Mao Tsetung Thought guides us in conquering Nature," *Peking Review*, 21 November 1969.
5. The general conception developed in the following paragraphs (but not its specific details) owes much to various writings by Marx, as well as to Scheler's *Die Wissensformen und die Gesellschaft*, Horkheimer's *Anfänge der bürgerlichen Geschichtsphilosophie*, and Lukács's *History and Class Consciousness*.
6. *Philosophical Rudiments concerning Government and Society*, Ch. 1, in *English Works*, Vol. II; *De corpore politico, ibid.*, Vol. IV, pp. 120–121.
7. *Capital*, Vol. I, Part Eight.
8. *The New Organon*, in *Works*, Vol. IV, p. 115.
9. *Capital*, Vol. I, p. 177.
10. Cf. Habermas, *Toward a Rational Society*, pp. 57–61.
11. C. S. Lewis, *The Abolition of Man*, pp. 39–41.
12. *Einbahnstrasse*, p. 125.

LIST OF WORKS CITED

ADAMS, Robert P. "The Social Responsibilities of Science in *Utopia, New Atlantis,* and After." *Journal of the History of Ideas,* X (1949), 374–398.

ADORNO, Theodor W. "Marx est-il dépassé?" In: *Marx and Contemporary Scientific Thought.* Publications of the International Social Science Council, No. 13. The Hague, 1969.

AHLERS, Rolf. "Is Technology Intrinsically Repressive?" *Continuum,* VIII (1970), 111–122.

D'ALEMBERT, Jean le Rond. *Preliminary Discourse to the Encyclopedia of Diderot.* Translated by R. N. Schwab and W. Rex. Indianapolis: Bobbs-Merrill, n.d.

ANDERSON, Fulton H. *Francis Bacon: His Career and His Thought.* University of Southern California Press, 1962.

——. *The Philosophy of Francis Bacon.* University of Chicago Press, 1948.

AQUINAS, St. Thomas. *Summa Theologica,* 22 volumes. Translated by the Fathers of the English Dominican Province. London: Burns, Oates and Washbourne, 1920–24.

ARENDT, Hannah. *The Human Condition.* University of Chicago Press, 1958.

ASSELIN, David T., S. J. "The Notion of Dominion in Genesis 1–3." *Catholic Biblical Quarterly,* XVI (1954), 277-294.

AXELOS, Kostas. *Marx, penseur de la technique: De l'aliénation de l'homme à la conquête du monde.* 3rd edition. Paris: Les Éditions de Minuit, 1969.

BACHELARD, Gaston. *La formation de l'esprit scientifique.* 5th edition. Paris: J. Vrin, 1967.

BACON, Francis. *Essays, and Wisdom of the Ancients.* Boston: Little, Brown and Co., 1891.

------. *New Atlantis.* Edited by A. B. Gough. Oxford: Clarendon Press, 1915.

------. *New Atlantis.* Edited by G. C. Moore Smith. Cambridge University Press, 1900.

------. *The Philosophy of Francis Bacon.* Edited and translated by B. Farrington. University of Chicago Press, 1966.

------. *The Works of Francis Bacon.* 7 volumes. Edited by J. Spedding, R. L. Ellis, and D. D. Heath. Many editions with identical volume and page numbering.

BAHR, Hans-Dieter. *Kritik der "Politischen Technologie."* Frankfurt: Europäische Verlagsanstalt, 1970.

BENJAMIN, Walter. *Einbahnstrasse.* Frankfurt: Suhrkamp, 1955.

BERGMANN, Joachim. "Technologische Rationalität und spätkapitalistische Ökonomie." In: *Antworten auf Herbert Marcuse,* ed. J. Habermas. Frankfurt: Suhrkamp, 1968.

BERNAL, John D. *The Social Function of Science.* London: George Routledge and Sons, 1939.

BLACK, John. *The Dominion of Man.* Edinburgh: University Press, 1970.

BOGUSLAW, Robert. *The New Utopians: A Study of System Design and Social Change.* Englewood Cliffs, N.J.: Prentice-Hall, 1965.

BORKENAU, Franz. *Der Übergang vom feudalen zum bürgerlichen Weltbild: Studien zur Geschichte der Philosophie der Manufakturperiod.* Paris: Félix Alcan, 1934.

BRECHT, Bertolt. *The Life of Galileo.* Translated by D. I. Vesey. London: Methuen, 1968.

------. *Tales from the Calendar.* Translated by Y. Kapp. London: Methuen, 1961.

BROWN, Norman O. *Life against Death.* New York: Vintage Books, n.d.

BUTLER, E. M. *The Fortunes of Faust.* Cambridge University Press, 1952.

------. *The Myth of the Magus.* Cambridge University Press, 1948.

CASSIRER, Ernst. *Das Erkenntnisproblem in der Philosophie und Wissenschaft der neueren Zeit.* 3 volumes. 2nd edition. Berlin: Verlag B. Cassirer, 1920–22.

------. *Idee und Gestalt.* Berlin: Verlag B. Cassirer, 1921.

------. "Mathematical mysticism and mathematical science." In:

Galileo, Man of Science, ed. E. McMullin. New York: Basic Books, 1967.

———. *The Philosophy of the Enlightenment.* Translated by F. C. A. Koelln and J. P. Pettegrove. Princeton University Press, 1951.

CASSIRER, Ernst, P. O. Kristeller, and J. H. Randall, eds. *The Renaissance Philosophy of Man.* Chicago: University Press, 1948.

CUVIER, Georges. *Éloges lus dans les séances publiques de l'Institut Royal de France.* 3 volumes. Paris: Levrault, 1819.

DEDIJER, Steven. "The Science of Science: A Programme and a Plea." *Minerva,* IV (1966), 489–504.

DESCARTES, René. *Philosophical Works.* Translated by E. Haldane and G. Ross. 2 volumes. New York: Dover Books, 1955.

DEWEY, John. *Experience and Nature.* London: George Allen and Unwin, 1929.

Doctrine Saint-Simonienne: Exposition. Paris: Librairie Nouvelle, 1854.

DUBOS, René. *The Dreams of Reason.* Columbia University Press, 1961.

———. *Reason Awake.* Columbia University Press, 1970.

DUPUY, Maurice. *La philosophie de Max Scheler.* 2 volumes. Paris: Presses Universitaires de France, 1959.

ELIADE, Mircea. *The Forge and the Crucible.* Translated by S. Corrin. London: Rider and Co., 1962.

ELLUL, Jacques. *The Technological Society.* Translated by J. Wilkinson. New York: Vintage Books, 1967.

ENGELS, Frederick. *Anti-Dühring.* Moscow: Foreign Languages Publishing House, 1954.

EURICH, Nell. *Science in Utopia.* Harvard University Press, 1967.

FALLOT, Jean. *Marx et la machinisme.* Paris: Éditions Cujas, 1966.

FEUER, Lewis. *The Scientific Intellectual.* New York: Basic Books, 1963.

FEUERBACH, Ludwig. *Sämtliche Werke.* Edited by W. Bolin and F. Jodl. 13 volumes. Reprint edition, Stuttgart, 1959–64.

FONTENELLE, Bernard le Bouvier de. *Oeuvres diverses.* 3 volumes. La Haye: Gosse and Neaulme, 1729.

FORBES, R. J. *The Conquest of Nature: Technology and its Consequences.* New York: Praeger, 1968.

FREUD, Sigmund. *Civilization and its Discontents.* Translated by J. Riviere. London: Hogarth Press, 1946.

FRYE, Northrop. "Varieties of Literary Utopias." In: *Utopias and Utopian Thought,* ed. F. Manuel. Boston: Beacon Press, 1967.

GAY, Peter. *The Enlightenment: An Interpretation.* 2 volumes. New York: Knopf, 1966–69.

GIDE, André. *Lafcadio's Adventures (Les Caves du Vatican).* Translated by D. Bussy. New York: Knopf, 1928.

GLACKEN, Clarence J. *Traces on the Rhodian Shore: Nature and Culture in Western Thought from Ancient Times to the End of the Eighteenth Century.* University of California Press, 1967.

GRAHAM, Loren. *The Soviet Academy of Sciences and the Communist Party, 1927–1932.* Princeton University Press, 1967.

GRANGER, Gilles-Gaston. *La mathématique sociale du Marquis de Condorcet.* Paris: Presses Universitaires de France, 1956.

GROSSMAN, Henryk. "Die gesellschaftlichen Grundlagen der mechanistischen Philosophie und die Manufaktur." *Zeitschrift für Sozialforschung,* IV (1935), 161–231.

HABERMAS, Jürgen. *Knowledge and Human Interests.* Translated by J. Shapiro. Boston: Beacon Press, 1971.

——. *Toward a Rational Society.* Translated by J. Shapiro. Boston: Beacon Press, 1970.

HALDANE, J. B. S. *Daedalus, or Science and the Future.* London: Kegan Paul, 1924.

HANKINS, Thomas L. *Jean d'Alembert: Science and the Enlightenment.* Oxford: Clarendon Press, 1970.

HAZARD, Paul. *The European Mind.* Translated by J. L. May. London: Hoddis and Carter, 1953.

HEGEL, G. W. F. *Lectures on the Philosophy of History.* Translated by J. Sibree. London: George Bell and Sons, 1894.

——. *Phenomenology of the Spirit.* Translated by J. B. Baillie. 2nd edition. London: George Allen and Unwin, 1949.

——. *Philosophy of Nature* (Part II of the *Encyclopedia of the Philosophical Sciences*). Translated by M. J. Petry. 3 volumes. London: George Allen and Unwin, 1970.

HEIDEGGER, Martin. *Holzwege.* Frankfurt: V. Klostermann, 1950.

HEISENBERG, Werner. *The Physicist's Conception of Nature.* Translated by A. J. Pomerans. London: Hutchinson, 1958.

——. *Physics and Philosophy.* New York: Harper and Row, 1962.

HOBBES, Thomas. *English Works.* 11 volumes. Edited by W. Molesworth. London: Bohn, 1839–45.

HOFFMEISTER, Johannes. *Goethe und der deutsche Idealismus.* Leipzig: F. Meiner, 1932.

HORKHEIMER, Max. *Anfänge der bürgerlichen Geschichtsphilosophie.* Stuttgart: W. Kohlhammer, 1930.

———. *Eclipse of Reason.* Columbia University Press, 1947.

———. *Kritische Theorie.* 2 volumes. Edited by A. Schmidt. Frankfurt: S. Fischer, 1968.

———. *Zur Kritik der instrumentellen Vernunft.* Edited by A. Schmidt. Frankfurt: S. Fischer, 1967.

———, and Theodor W. Adorno. *Dialektik der Aufklärung.* Frankfurt: S. Fischer, 1969.

———. *Sociologica II: Reden und Vorträge.* Frankfurt: Europäische Verlagsanstalt, 1962.

HOROWITZ, Irving L., ed. *The Rise and Fall of Project Camelot.* Massachusetts Institute of Technology Press, 1967.

HUSSERL, Edmund. *Ideen zu einer reinen Phänomenologie und phänomenologischen Philosophie,* Vol. II. Edited by M. Biemel. The Hague: M. Nijhoff, 1952.

———. *Die Krisis der europäischen Wissenschaften und die transzendentale Phänomenologie.* Edited by Walter Biemel. 2nd edition. The Hague: M. Nijhoff, 1962. Translated by David Carr as *The Crisis of European Sciences and Transcendental Phenomenology.* Evanston, Ill.: Northwestern University Press, 1970.

HUXLEY, Aldous. "Commentary." In: *The Technological Order,* ed. C. Stover. Detroit: Wayne State University Press, 1963.

JACOBY, Russell. "Reversals and Lost Meanings." In: *Critical Interruptions,* ed. P. Breines. New York: Herder and Herder, 1970.

JANSON, H. W. *Apes and Ape-Lore in the Middle Ages and the Renaissance.* London: The Warburg Institute, 1952.

JORAVSKY, David. *Soviet Marxism and Natural Science, 1917–1932.* Columbia University Press, 1961.

JUNG, Carl G. *Alchemical Studies (Collected Works,* Vol. XIII). Translated by R. F. C. Hull. Princeton University Press, 1967.

———. *Psychology and Alchemy (Collected Works,* Vol. XII). Translated by R. F. C. Hull. 2nd edition. Princeton University Press, 1968.

KATEB, George. *Utopia and its Enemies.* New York: The Free Press, 1963.

KENISTON, Kenneth. "Does Human Nature Change in a Technological Revolution?" *The New York Times,* 6 January 1969.

KENNINGTON, Richard. "René Descartes." In: *History of Political Philosophy,* eds. L. Strauss and J. Cropsey. University of Chicago Press, 1963.

KNIGHT, Isabel F. *The Geometric Spirit: The Abbé de Condillac and the French Enlightenment.* Yale University Press, 1968.

KOJÈVE, Alexandre. "L'Origine chrétienne de la science moderne." In: *Mélanges Alexandre Koyré.* Vol. II. Paris: Hermann, 1964.

KOYRÉ, Alexandre. *Études Galiléennes.* 3 volumes. Paris: Hermann, 1939.

——. *Études d'Histoire de la pensée scientifique.* Paris: Presses Universitaires de France, 1966.

——. *From the Closed World to the Infinite Universe.* Baltimore: Johns Hopkins University Press, 1957.

——. *Mathematics and Measurement: Essays in Scientific Revolution.* London: Chapman and Hall, 1968.

——. *Newtonian Studies.* London: Chapman and Hall, 1965.

LANDGREBE, Ludwig. *Major Problems in Contemporary European Philosophy.* Translated by K. F. Reinhardt. New York: Ungar, 1966.

LEISS, William. "The Domination of Nature." Ph.D. thesis. University of California at San Diego, 1969.

——. "Husserl and the Mastery of Nature." *Telos,* No. 5 (1970), pp. 82–97.

——. "Husserl's *Crisis of European Sciences." Telos,* No. 8 (1971).

——. "Max Scheler's Concept of Herrschaftswissen." *The Philosophical Forum,* Vol. II, No. 3 (Spring, 1971), pp. 316–331. In Spanish translation: *Dialogos,* No. 16 (1969), 29–52.

——. "The Social Consequences of Technological Progress: Critical Comments on Recent Theories." *Canadian Public Administration,* XIII (1970), 246–262.

——. "Utopia and Technology: Reflections on the Conquest of Nature." *International Social Science Journal,* XXII (1970), 576–588.

LENOBLE, Robert. *Mersenne ou la naissance du mécanisme.* Paris: J. Vrin, 1943.

LEWIS, C. S. *The Abolition of Man.* London: Geoffrey Bles, new edition, 1946.

LUKÁCS, Georg. *History and Class Consciousness.* Translated by R. Livingstone. London: Merlin Press, 1971.

MACPHERSON, C. B. *The Political Theory of Possessive Individualism.* Oxford University Press, 1962.

MANUEL, Frank. *A Portrait of Isaac Newton.* Harvard University Press, 1968.

——. *The Prophets of Paris.* Harvard University Press, 1962.

MARCUSE, Herbert. *Eros and Civilization.* 2nd edition. Boston: Beacon Press, 1966.

———. *An Essay on Liberation*. Boston: Beacon Press, 1969.

———. *Five Lectures*. With translations by J. Shapiro and S. Weber. Boston: Beacon Press, 1970.

———. *Negations: Essays in Critical Theory*. With translations by J. Shapiro. Boston: Beacon Press, 1968.

———. "The Obsolescence of Marxism?" In: *Marx and the Western World*, ed. N. Lobkowicz. University of Notre Dame Press, 1967.

———. *One-Dimensional Man*. Boston: Beacon Press, 1964.

———. "On Science and Phenomenology." In: *Boston Studies in the Philosophy of Science*, Vol. II, eds. R. Cohen and M. Wartofsky. New York: Humanities Press, 1965.

———. "Re-examination of the Concept of Revolution." In: *Marx and Contemporary Scientific Thought*. Publications of the International Social Science Council, No. 13. The Hague, 1969.

———. "The Responsibility of Science." In: *The Responsibility of Power*, eds. L. Krieger and F. Stern. New York: Doubleday, 1967.

———. "Some Social Implications of Modern Technology." *Studies in Philosophy and Social Science*, IX (1941), 414–439.

———.*Soviet Marxism*. New York: Vintage Books, 1961.

MARSAK, Leonard. *Bernard de Fontenelle: The Idea of Science in the French Enlightenment*. Transactions of the American Philosophical Society (New Series), Vol. 49, Part 7. Philadelphia, 1959.

MARX, Karl. *Capital*. 3 volumes. Moscow: Foreign Languages Publishing House, 1961.

———. *Grundrisse der Kritik der politischen Ökonomie*. Berlin: Dietz Verlag, 1953.

———, and Frederick Engels. *The German Ideology*. Moscow: Foreign Languages Publishing House, 1964.

MATSON, Floyd. *The Broken Image: Man, Science and Society*. New York: Braziller, 1964.

MIDWEST RESEARCH INSTITUTE. "Assessment of Ecological Effects of Extensive or Repeated Use of Herbicides." Washington, D.C.: Department of Commerce, 1967.

MOORE, Barrington, Jr. "American nightmare: Why we fear peasants in revolt." *The Nation*, 26 September 1966, pp. 271–274.

———. "The Society Nobody Wants." In: *The Critical Spirit: Essays in Honor of Herbert Marcuse*, eds. K. Wolff and B. Moore. Boston: Beacon Press, 1967.

MORE, St. Thomas. *Complete Works*. Edited by E. Surtz and J. H. Hexter. In progress. Yale University Press, 1963–.

MUMFORD, Lewis. *The Myth of the Machine*. New York: Harcourt, Brace and World, 1966.

——. "Utopia, the City, and the Machine." In: *Utopias and Utopian Thought,* ed. F. Manuel. Boston: Beacon Press, 1967.

MURPHY, Earl. *Governing Nature.* Chicago: Quadrangle Books, 1967.

NEEDHAM, Joseph. *The Grand Titration: Science and Society in East and West.* London: George Allen and Unwin, 1969.

NIETZSCHE, Friedrich. *The Will to Power.* Edited and translated by W. Kaufmann. New York: Random House, 1967.

OFFE, Claus. "Technik und Eindimensionalität." In: *Antworten auf Herbert Marcuse,* ed. J. Habermas. Frankfurt: Suhrkamp, 1968.

PELIKAN, Jaroslav. "Cosmos and Creation: Science and Theology in Reformation Thought." In: *Proceedings of the American Philosophical Society,* Vol. 105, No. 5 (1961). Philadelphia, 1961.

PRIOR, Moody. "Bacon's Man of Science." In: *The Rise of Science in Relation to Society,* ed. L. Marsak. New York: Macmillan, 1964.

PURVER, Margery. *The Royal Society: Concept and Creation.* London: Routledge and Kegan Paul, 1967.

RABB, Theodore. "Puritanism and the Rise of Experimental Science in England." In: *The Rise of Science in Relation to Society,* ed. L. Marsak. New York: Macmillan, 1964.

RANDALL, John H. *Aristotle.* Columbia University Press, 1960.

RANSOM, David. "The Berkeley Mafia and the Indonesian Massacre." *Ramparts,* Vol. IV, No. 4 (October 1970).

ROSSI, Paolo. *Francis Bacon: From Magic to Science.* Translated by S. Rabinovitch. London: Routledge, 1968.

——. *Philosophy, Technology and the Arts in the Early Modern Era.* Translated by S. Attanasio. New York: Harper and Row, 1970.

RUSSELL, Bertrand. *Icarus, or the Future of Science.* London: Kegan Paul, 1924.

SCHELER, Max. *Schriften aus dem Nachlass,* Band I (*Gesammelte Werke,* Band 10). 2nd edition. Bern: Francke Verlag, 1957.

——. *Schriften zur Soziologie und Weltanschauungslehre (Gesammelte Werke,* Band 6). 2nd edition. Bern: Francke Verlag, 1963.

——. *Die Wissensformen und die Gesellschaft (Gesammelte Werke,* Band 8). 2nd edition. Bern: Francke Verlag, 1960.

SCHMIDT, Alfred. *Der Begriff der Natur in der Lehre von Marx.* Frankfurt: Europäische Verlagsanstalt, 1962. English translation: *The Concept of Nature in Marx.* London: New Left Books (forthcoming).

SCHWONKE, Martin. *Vom Staatsroman zur Science Fiction.* Stuttgart: Enke, 1957.

SEARS, Paul B. "Utopia and the Living Landscape." In: *Utopias and Utopian Thought,* ed. F. Manuel. Boston: Beacon Press, 1967.

SHAPIRO, Jeremy. "One-Dimensionality: The Universal Semiotic of Technological Experience." In: *Critical Interruptions,* ed. P. Breines. New York: Herder and Herder, 1970.

SIDGWICK, Peter. "Natural Science and Human Theory." In: *Socialist Register 1966,* eds. R. Miliband and J. Saville. New York: Monthly Review Press, 1966.

SIMON, Yves. *The Philosophy of Democratic Government.* University of Chicago Press, 1951.

SKINNER, B. F. *Walden Two.* New York: Macmillan, 1962.

STAUDE, John R. *Max Scheler.* New York: The Free Press, 1967.

THORNDIKE, Lynn. *A History of Magic and Experimental Science.* 8 volumes. Columbia University Press, 1923–1958.

UNESCO. Intergovernmental Conference of Experts on the Scientific Basis for Rational Use and Conservation of the Resources of the Biosphere. "Final Report." UNESCO Document SC/MD/9 (1969).

WALKER, Daniel P. *Spiritual and Demonic Magic from Ficino to Campanella.* London: Warburg Institute, 1958.

WARTOFSKY, Marx W. *Conceptual Foundations of Scientific Thought.* New York: Macmillan, 1968.

WEISZÄCKER, Carl F. *The History of Nature.* Translated by F. D. Wieck. University of Chicago Press, 1949.

——. *The World View of Physics.* Translated by M. Greene. University of Chicago Press, 1952.

WHITE, Lynn, Jr. *Machina ex Deo: Essays in the Dynamism of Western Culture.* Massachusetts Institute of Technology Press, 1968.

WOLF, Eric and Joseph Jorgensen. "Anthropology on the Warpath in Thailand." *The New York Review of Books,* Vol. XV, No. 9 (November 19, 1970), pp. 26–35.

YATES, Frances. *Giordano Bruno and the Hermetic Tradition.* London: Routledge and Kegan Paul, 1964.

ZILSEL, Edgar. "Copernicus and Mechanics," "The Genesis of the Concept of Scientific Progress," and "The Origins of Gilbert's Scientific Method." In: *Roots of Scientific Thought,* eds. P. Wiener and A. Noland. New York: Basic Books, 1957.

——. "The Genesis of the Concept of Physical Law." *Philosophical Review,* LI (1942), 245–279.

——. "The Sociological Roots of Science." *American Journal of Sociology,* XLVII (1942), 544–562.

INDEX*

Absolutization, *see* Scientific methodology
Abstraction, modes of, 110–111, 136, 141
Adam, 51, 53, 188
Adams, Robert P., 216 (note 25)
Adorno, Theodor, 11, 150, 200
Ahlers, Rolf, 200
Alchemy, 36, *40–44,* 47, 48, 51, 60, 74, 76, 80
d'Alembert, 45, 78
Anderson, Fulton, 215 (notes 1 and 4)
Andreae, Johann, 68
Aquinas, St. Thomas, 31–32
Archaic impulses, 27–28, 40, 42, 44, 50
Arendt, Hannah, 32
Aristotle, 14, 87, 107–108, 109, 120, 149, 150, 182, 183
Asselin, David, 214 (note 6)
Astrology, 36, 38, 41, 74, 75
Autonomy, *see* Individual and society
Axelos, Kostas, 217 (note 15)

Bachelard, Gaston, 216 (note 21)
Bacon, Francis, 14, 21, 25, 32, 35, 44, *45–71,* 79, 80, 87, 94, 96, 132, 134, 138, 139, 140, 177, 185, 186, 188, 189, 191, 192, 193, 196
  conceptualization of mastery over nature, 57–60
  interpretation of ancient myths, 3–4, 6, 11, 69
  interpretation of Christian doctrine, 48–55
  later evaluations of, 45–48
  New Atlantis contrasted with More's *Utopia,* 61–70
Bahr, Hans–Dieter, 200
Benjamin, Walter, 198
Bensalem, 64, 66, 67, 68, 70, 198
Bergmann, Joachim, 201
Bernal, J. D., 4, 95
Black, John, 214 (note 4)
Boguslaw, Robert, 14, 17
Borkenau, Franz, 90–92
Brecht, 3, 45
Brown, Norman O., 60

---

* I am indebted to Jo Roberts and Marilyn Lawrence for assistance with the preparation of the manuscript and the index. W.L.

233

238